From Science to Seapower

A Roadmap for S&T Revitalization

From Science to Seapower

A Roadmap for S&T Revitalization

Robert A. Kavetsky
Michael L. Marshall
Davinder K. Anand

Center for Energetic Concepts Development Series

CALCE
EPSC Press

University of Maryland, College Park, Maryland

Endorsements

"This timely analysis should be of great interest to those who share concern over maintaining U. S. dominance in the science and technology fields. World leadership in nurturing and developing science and technology expertise has remarkably contributed to U. S. superiority in advanced defense weaponry and overall competence in systems engineering."

Vice Admiral Michael Colley, United States Navy, is a retired submariner who has long championed the importance of U. S. science and technology superiority. He was Director of the Division of Mathematics and Science at the Naval Academy and Deputy Commander of the U. S. Strategic Command. He also commanded the submarine force in the Pacific theater during the first Gulf war.

"This book addresses a systemic problem that has developed over the past few decades. The book details the history of the development of the problem and lays out a rational set of actions required to begin to solve the problem. It addresses an issue currently facing the whole country, not just the DOD. For those interested in science and technology it is a must read."

Dr. James Colvard is currently a Fellow of the National Academy of Public Administration and a visiting professor at Virginia Tech. During his long and distinguished career of federal service, Dr. Colvard held many important positions including Technical Director of the Naval Surface Weapons Center, Deputy Chief of the Naval Material Command, Director of Civilian Personnel Policy and EEO of the Navy and Deputy Director of the Office of Personnel Management during the

Reagan Administration. He also served as associate director of the Johns Hopkins University Applied Physics Laboratory.

"A wide-ranging synthesis of the current literature, *From Science to Sea Power* forcefully demonstrates the need to maintain a robust science and technology effort within the DOD."

Dr. George Dieter is a member of the National Academy of Engineering, a former Dean of the College of Engineering and now Glen L. Martin Professor of Engineering at the University of Maryland, College Park.

"From *Science to Sea Power* addresses aspects of a potential crisis in safeguarding continued superiority of our naval technology in a world of rapidly changing threats to our national security. Finding the right new balance point in providing resources for technology unique to our naval supremacy compared with dependence on what is available from surging new global commercial sources has to be an absorbing task for naval technical leadership. *From Science to Sea Power* brings insight, data and commitment to this important question."

Dr. Millard Firebaugh, Rear Admiral United States Navy (ret), was Chief Engineer and Deputy Commander for Design and Engineering at the Naval Sea Systems Command. Since retiring from the Navy he has held executive positions involved in advanced technology, including seven and a half years at General Dynamics, Electric Boat Corporation, where he was Vice President Innovation and Chief Engineer. Currently, he is President and Chief Operating Office of SatCon Technology Corp. Dr. Firebaugh is a member of the National Academy of Engineering.

"This book is ringing a bell and I hope it is echoing within the Pentagon's corridors. It argues persuasively for investment in an increasingly competitive, globalized and complex technology future. And it implies, while there is no power like that of the American marketplace, that certain technologies will not be fielded if their development is left simply to profit motive. The DOD lab system, cheaper than alternatives and better, by a million miles, in many technological situations, needs a few top level champions ... now. I

recommend this book to federal administrators in and outside of the **Defense Department, students, members of the Congressional enterprise and to leaders in the American technology industry."**

Paul G. Gaffney II, Vice Admiral United States Navy (ret.), is currently president of Monmouth University in West Long Branch, New Jersey. Before retiring, Gaffney had a long and distinguished military career during which he held numerous important positions including that of president of the National Defense University in Washington, D.C., and that of Chief of Naval Research in which position he was responsible for the Navy's science and technology investment.

"American preeminence in science and technology has been fundamental to both the security of our nation and the vitality of our economy for many years. *From Science to Sea Power* **identifies growing challenges to that preeminence that demand our collective attention."**

Mike Hayes, Brigadier General United State Marine Corps (Ret.) served on active duty for 33 years, commanding Marines during two tours in Vietnam, as well as at Quantico, Parris Island, Camp Pendleton, Panama and Okinawa. He is currently the Director of Military and Federal Affairs for the State of Maryland.

"A succinct and convincing overview of the crucial workforce and investment issues faced in the Department of Defense science and technology enterprise."

Dr. Robert C. Kolb is a former Executive Director of the Space and Naval Warfare Systems Center, San Diego, California. During his 37 years in Navy Laboratories, he held many leadership roles in Science and Technology including serving as Chair of the DOD Information Systems Technology Reliance Panel.

"This book is an important contribution to the discussion of the future of DOD laboratories. While addressing primarily Department of Navy laboratories, the problems, issues, frustrations,

as well as proposed solutions apply equally to all DOD S&T activities."

Dr. Donald Lamberson, Major General United States Air Force (ret), has spent the majority of his active duty and post-retirement career in USAF Laboratory and Acquisition positions. At the time of his retirement in 1989 he was Assistant Deputy, Office of the Assistant Secretary of the Air Force for Acquisition.

"By far, the best document I've read for both justifying an in-house government R&D capability and laying out the challenges of developing and sustaining a government R&D cadre. Terms in the glossary can be changed from Navy to DOD, Army, Air Force, NASA or DOE and the conclusions would be the same. This book also helps industry better comprehend why it needs a first-rate in-house government scientific and engineering community - to assess the value and credibility of technological options and enable industry to better support the warfighter. The conclusions with regard to revitalization of the national security research workforce apply across-the-board both within and outside the government. The authors make an eloquent case that both industry and the government benefit from public-private technological partnership. This document offers strong, logical arguments, well-documented findings and thoughtful conclusions. No problem worrying about return-on-investment - the book will pay for itself."

Dr. Malcolm R. O'Neill, Lieutenant General United States Army (ret), is currently Vice President and Chief Technical Officer of Lockheed Martin Corporation. During his military career, he held numerous prominent positions including Director of the Ballistic Missile Defense Organization; Deputy Director, Strategic Defense Initiative Organization; and Commander, Army Laboratory Command. A native of Chicago, Illinois, O'Neill received his B.S. in physics from De Paul University and his M.S. and Ph.D. in physics from Rice University.

Preface

Two months before its use, the thermobaric warhead used with great success in Afghanistan had been little more than an idea. A remarkably rapid transition from concept to reality—a process that usually takes years—occurred in large part because of the work done by explosives scientists at the Indian Head, Maryland Division of the Naval Surface Warfare Center (NSWC). Some ten years earlier, they had developed the formula for an experimental explosive they called PBXIH-135. Although initial tests proved promising, the explosive had never been transitioned into an actual weapon.

Fortunately, however, when the emergency arose, the Departments of Defense (DOD) and the Department of the Navy (DON) had "in-house" not only the seeds for the critical capability, but also the best explosives experts in the world. In fact, scientists at Indian Head developed almost 90 percent of the explosives used in American weapons today. Altogether, in 67 days, some 100 employees at the Division completed a process the DON's in-house laboratories and centers have been performing for six decades: the transition from science to sea power—or in this case, air power.

Although the thermobaric warhead is only one example of hundreds of weapons and other materiel the DOD's in-house laboratories and centers have conceived and produced, these technical institutions remain an under-appreciated—in fact sometimes unknown, at least by the general public—element of U.S. national security. Since the end of World War II, the scientists and engineers (S&Es) at these centers have researched, developed, tested, and evaluated the technology that evolves into the weapons, weapons systems, apparatus, supplies, and equipment that underpin the most sophisticated and effective defense systems the world has ever known.

Rear Admiral Jay Cohen, the Chief of Naval Research echoed these observations in his congressional testimony of March 2003: *"The effectiveness of the war-fighting systems employed by the Navy and Marine Corps of the Future depends as much on investment in these* dedicated, capable civil servants as it does on the size of the science and technology budget itself. The past decade's frequent downsizings, coupled with the declining number of American students-particularly women and minorities, pursuing mathematics, engineering and physical science degrees-has left us with a dwindling pool of scientists and engineers available to become the next generation of researchers. This situation jeopardizes our ability to perform essential research in support of, ultimately, Sailors and Marines."

This book describes the nature and extent of these and some other major threats to the vitality of DOD's in-house S&T enterprise, and offers recommendations that could help either to reverse the most disturbing trends or to address some of the underlying causes of long-term problems. It synthesizes a wide-ranging array of literature on a variety of workforce, funding, and science and technology (S&T) innovation topics. Although the focus is primarily on DON laboratories and centers, many of the sources discussed and conclusions drawn apply DOD-wide.

Of course, efficiencies aimed at producing savings should always be sought, the end of the Cold War justified some defense drawdown, and the DOD has always contracted out most of its R&D to the private sector. But a quest aimed at cost-cutting, combined with dubious articles of faith concerning the value of privatization should not and, indeed, cannot be defied to the point of destroying the in-house core competencies that concrete evidence has time and again demonstrated to be at the heart of our national defense. These core competencies should be protected and maintained at a cutting-edge state.

Executive Summary

The defense technology base, a coalition of public and private partners working to ensure U.S. military preeminence, grew from America's experience in World War II. This coalition successfully prosecuted a massive effort, as academic, industrial, and Federal Government research and development centers worked together to support the war. The success of this partnership led to increasing reliance on federally funded research and development in the post-War period. Indeed, throughout the Cold War, a technological superiority sufficient to overcome numerical disadvantages in men and materiel was the bedrock of deterrence.

Today, the vitality of the defense science and technology (S&T) effort is threatened. Funding necessary to execute long-term research and development (R&D) continues to dwindle. In addition, as a generation of highly capable scientists and engineers (S&Es) nears retirement, there are fewer and fewer young technologists to replace them, and the Department of Defense (DOD) in-house centers have been hamstrung in their efforts to provide challenging work, state of the art facilities, and even public recognition that could attract a new generation of top-quality S&Es. All of this means not only that the in-house talent pool grows shallower, but also that invaluable corporate memory or "deep smarts" that retirees possess will not be transferred to the next generation, and thus lost forever.

There are numerous causes contributing to this situation. A decade of downsizing, consolidations, and closures, an outsourcing fad which has resulted in costs for new military systems spiraling out-of-control, and short-sighted cutbacks in discretionary funding have strained these centers' ability to provide cutting-edge work and otherwise engage proactively in employment of premier talent. Similarly, many of the reductions, aimed at improving short-term efficiency, have crippled the centers' capacity to execute work that is by its nature long-term. In the Department of the Navy (DON), these realities exist in tandem with military construction (MILCON) and Working Capital Fund requirements, which obstruct purchases of modern S&T equipment. And centralization across the DOD, aimed at enhancing accountability, in many cases actually impedes system responsiveness in the S&T community.

The bow wave of senior S&Es moving through the system towards retirement, along with continued funding cuts and outsourcing, not only imperil the centers' ability to carry out the core competencies that defense analysts have for decades considered essential, but also jeopardize the technological superiority upon which America's national security depends. Basic and applied research are the seeds from which future military capabilities grow. And while all kinds of trend data suggest the U.S. will not have enough qualified S&Es to perform such research, other countries are training exponentially increasing numbers of them, offering attractive opportunities for students while building increasingly sophisticated technical infrastructures. College enrollment and post-graduate degrees conferred in those countries have skyrocketed, while the numbers in the U.S. remain the same or are decreasing. In addition, off-shoring practices of many U.S. companies further contribute to technical infrastructure development elsewhere. In short, the science and engineering (S&E) talent pool for other countries is getting considerably deeper.

The negative effects of these trends—which are still in their early stages—are already obvious. For example, various indicators of technical outputs (papers published, citations of those papers, patent citations and applications, patents granted, patent citations to S&E articles) show other countries clearly gaining on the U.S., some of them rapidly. Defense commentators who focus on simply a snapshot of the current situation are unjustifiably confident. Just as a snapshot taken half way through a horse race is not a good predictor of which horse will ultimately win the race, a snapshot of today's global trends will not reveal where the U.S. may be headed in the future. Further analysis of the global picture also indicates America's current leadership position stems from dominance in biomedical research and clinical medicine, which are certainly important, but less so to the DOD.

In fact, a shift in Federal funding toward these medically related life sciences mirrors the shift away from funding for the basic and applied sciences, and has ominous implications for national defense and other areas. Overemphasis on medically related fields has occurred at the expense of fields focused on research necessary for the DOD. Many disciplines suffering cutbacks depend most directly on Federal funding, and are imperative to national defense and other areas because of the increasingly interdisciplinary nature of scientific and technological advances.

Such developments are all the more unsettling in light of what is occurring in many other countries. Trends in two fields, nanotechnology and energetic materials (EMs), show how some countries have resourcefully focused their R&D investments, while the U.S. has shortsightedly under funded areas crucial to the smaller, cheaper, more lethal smart weapons on which the DOD appears to be focusing.

Intertwined with these troubling workforce and funding reductions is a dramatically altered environment that demands adjustments in managing the transition from science to sea power. Post-Cold War threats and changes in the way innovation occurs have combined to render obsolete many of the methods by which the DOD has traditionally overseen this process. New "innovation networks" and "technology generation networks," new types of cooperative agreements, and interdisciplinary collaboration all indicate that today's S&Es need more interdisciplinary education, more total education, and more continuing education opportunities than ever before. The scientists and engineers who work in S&T in the future will need to have PhD-level credentials to sit at the global S&T table.

Just when new workforce paradigms are needed, however, influences operating in the current environment militate against creating them, and in fact further enlarge the circle of negative effects. For one, work at the labs and centers is not recognized and awarded by means such as membership in national academies. Two, the new National Security Personnel System (NSPS) is headed towards a one-size-fits-all arrangement that will not have the flexibility needed for the S&T community. Further, center directors do not have the flexibility or institutional mechanisms to incentivize high performers and thereby counteract freezes, downsizing, outsourcing, and centralization. These factors, along with others, have combined to affect the military component of the S&T enterprise, discouraging naval officers from pursuing careers in science and technology. The talent pipeline, then, is shrinking on both the military and civilian sides.

For over six decades, investments in DOD in-house S&T have conferred remarkable benefit-to-cost. The labs and centers have created budget savings and increased the reliability of weapons and warfare systems. They have created efficiencies in operations, manufacturing processes, and maintenance—the very things today's critics of S&T investment want emphasized. Their "yardstick" or "smart buyer" capability has offered objective advice about the work DOD contracts out to the private sector, and has time and again prevented costly acquisition errors.

Further, the labs and centers not only deliver basic and applied research that pays off in the long term, but also respond rapidly to threats and battlefield problems. Even after a decade of cutbacks, the DON warfare centers and corporate laboratory have provided timely, critical support to soldiers engaged in the war on terror. Thermobaric warheads used in the tunnels of Afghanistan and the Hellfire Missile, and the "Dragonshield" polymer coating that protects High-Mobility Multipurpose Wheeled Vehicles, are just a couple examples. The improvised explosive devices (IEDs) and similar devices used in Iraq have created an even more urgent demand for new technological breakthroughs with benefits difficult to reduce to a cost ratio: saving the lives of U.S. troops.

In fact, there are some indications that the DOD will once again turn to its in-house laboratories and centers. Many have observed that outsourcing itself carries burdensome costs. Further, the private sector is increasingly reluctant to invest in certain kinds of research the DOD must have. Also, globalization, and changes in academia and the defense industrial base, signify an escalating need for defense in-house laboratories. We believe that addressing these future challenges requires achieving four overarching goals:

- First, there must be a sufficiently large community of S&T "prospectors" who can participate as peers with colleagues throughout the global community, thereby ensuring that the fruits of the worldwide enterprise can also be applied.

- Second, the DON must develop and implement a human capital strategy for its in-house S&T community that will work on today's most important problems. The strategy should result in 4,000 new Ph.D.s by 2015.

- Third, S&T funding should be increased to 3 percent of overall DON total obligation authority, and it should be maintained at that level in constant, inflation-adjusted dollars through 2015 to ensure revitalization of the workforce and provide the intellectual capital base for the Navy-After-Next.

- Fourth, the DON must support and emphasize technical excellence, and appoint visionary civilian leaders who have the authority and responsibility for the technical output of their organizations.

To achieve these goals, the following recommendations are offered:

- Expand the Office of Naval Research (ONR) Global activities to include all the major players on the international S&T scene.

- The DON should bring on board 500 S&Es per year over the next ten years to pursue research and technology areas of critical importance to developing future military capabilities.

- Institute a DON S&T community expectation that career path to journeyman level requires a Ph.D. or equivalent education.

- Create a DOD S&T Academy, equivalent in prestige to the National Academies of Science and Engineering.

- Increase the annual DON S&T budget to three percent of DON Total Obligation Authority (TOA) and reallocate amongst community members (academia, in-house centers and industry) to ensure viability of each sector.

- Provide a $50 million laboratory and equipment-funding source in the DON S&T account to be focused on S&T frontiers.

- Launch an aggressive DON-wide program to ensure inter-generational transfer of corporate knowledge.

- Create an S&T Governance Council chaired by the Assistant Secretary of the Navy for Research, Development & Acquisition (ASN (RDA)) with membership that includes all major stakeholders, including the ONR, the Naval Research Laboratory (NRL), the Warfare/Systems Centers and University-Affiliated Research Centers (UARCs).

- Establish Senior Executive Service (SES) level Technical Director (TD) positions at the warfare/systems center division sites, and invest them with authority and responsibility for the entire technical output of the organization.

- Institute within military career paths a cadre of "Military Technology Officers".

As will be seen, these recommendations are closely aligned to those made in several recent studies carried by groups such as the National Academies. While they are neither unique, nor the only ones that could be made, it is considered that they are sufficient to provide a roadmap to help the DON rebuild its in-house technical institutions, namely its laboratories and centers. This is a task of great importance because these institutions, along with their private sector defense technology base partners, are vital to America's future national security.

Author Biographies

Bob Kavetsky is the Director of the N-STAR initiative at the Office of Naval Research, which is a Navy-wide effort aimed at reinvigorating the S&T community within the Navy's Warfare Centers. He received a BSME in 1975, a MSME in 1977, and a MEA in 1978, all from Catholic University. He was head of the Explosion Damage Branch, Program Manager for the 6.2 Undersea Warheads Program, and Program Manager for Undersea Weapons at the Naval Surface Warfare Center. At OPNAV in 1999-2000, he helped develop S&T programs for organic mine countermeasures and expeditionary logistics, and then at NSWC Indian Head created "Workforce 2010," a government, industry, and academic consortium focused on developing Indian Head's next generation workforce. He has authored a number of technical and S&T workforce-related policy and program publications for ASEE, ASME, AIAA, and other forums.

Michael L. Marshall has spent over 35 years in the Navy's R&D community. He has a B.S. and M.S. in physics from the University of North Carolina, Chapel Hill, and a J.D. from the University of Maryland. After working as a research physicist, program manager, and line manager at the Naval Surface Warfare Center, he served the Director of Navy Laboratories (DNL) as special assistant for science and technology, headed the DNL's corporate projects office, and served as Executive Secretary of the Navy Laboratory/Center Coordinating Group. Mr. Marshall then became Assistant to the Director of the Applied Research Laboratory at Penn State. He is the author of scores of technical publications, and also of reports published by the Naval Institute *Proceedings*, *Acquisition Review Quarterly*, Penn State, and *Defense Horizons*.

Davinder Anand is Emeritus Professor of Mechanical Engineering and Director of the Center for Energetic Concepts Development, both at the University of Maryland, College Park. He received his Ph.D. from

George Washington University in 1965, and from 1991-2002 chaired the Department of Mechanical Engineering at College Park. He has served as a Director of the Mechanical Systems Program at the National Science Foundation, and his research has been supported by NIH, NASA, DOE, DOD, and industry. He has lectured internationally, founded two high technology research companies (most recently Iktara and Associates, LLC), published three books and over one hundred and seventy papers, and has one patent. He is a Distinguished Alumnus of George Washington University, and was awarded the Outstanding and Superior Performance Award by the National Science Foundation. Dr. Anand is a Fellow of ASME and is listed in Who's Who in Engineering.

Acknowledgment

Although the authors had decided to make their case writing a book on Science to Seapower, it could not have been done without the help of many colleagues and their assistance is gratefully acknowledged. Eric Hazell did a masterful job at integrating the various contributors and provided the overall editing of the various chapters. Jim Short provided insightful comments and other collegial advice for the book, CECD and long-standing support for DOD scientists and engineers as well. Bob McGahern, Ernest McDuffie, and Gene Brown have, since the beginning of our revitalization effort, provided innovation, creativity, and backbone in building the N-STAR enterprise. Jennifer McGraw has shown us the power of a message well communicated, and a passion for all things from our warfare centers. Gary Hess supplied much of the DON data used and offered expert observations on several points regarding the book's overall direction. John Brough, Satyandra Gupta and Ron Kostoff provided useful material for Chapter 3. George Dieter and Ron Kostoff read the entire manuscript and made several good suggestions. Lise Crittenden helped in editing and making last minute changes and Ania Picard provided the final touches necessary for the manuscript printing.

Finally, and most personally, we gratefully acknowledge the enthusiastic support of our wives Lyn, Chris and Asha who unselfishly gave us the time required to focus the task at hand.

Acronyms

AFIT	Air Force Institute of Technology
AFRL	Air Force Research Laboratory
AGED	Pentagon Advisory Group on Electron Devices
AIA	Aerospace Industries Association
ASN (RDA)	Assistant Secretary of the Navy for Research, Development and Acquisition
ATD	Advanced Technology Development
BA	Budget Activity
BEST	Building Engineering and Scientific Talent
BtC	Benefit to Cost
BRAC	Base Realignment and Closure
CCMD	Civilian Community Management Division
CECD	Center for Energetic Concept Development
CNIC	Commander, Naval Installations
CNO	Chief of Naval Operations
CNR	Chief of Naval Research
CO	Commanding Officer
CRADA	Cooperative Research and Development Agreement
CRS	Congressional Research Service
CSP	Conference Support Program
CSRS	Civil Service Retirement System
CSS	Civil Service System
D&I	Discovery and Invention
DCPDS	Defense Civilian Personnel Data System
DDR&E	Director, Defense Research and Engineering
DMDC	Defense Manpower Data Center
DOD	Department of Defense
DOE	Department of Energy
DON	Department of the Navy
DSB	Defense Science Board
DTIC	Defense Technical Information Center
DURIP	Defense University Research Instrumentation Plan

Ems	Energetic Materials
EU	European Union
FFRDC	Federally Funded Research and Development Center
GAO	Government Accountability Office/Government Accounting Office
GDP	Gross Domestic Product
GWOT	Global War on Terrorism
HPE	High Performance Explosive
HPEC	High Performance Explosive Component
IDA	Institute for Defense Analysis
IED	Independent Exploratory Development
IEDs	Improvised Explosive Devices
ILIR	In-House Laboratory Independent Research
IR&D/IRAD	Independent Research and Development
MILCON	Military Construction
MREFC	Major Research Equipment and Facilities Construction
NAE	National Academy of Engineering
NAPA	National Academy of Public Administration
NASA	National Aeronautics and Space Administration
NAVAIR	Naval Air Systems Command
NAVSEA	Naval Sea Systems Command
NAWC	Naval Air Warfare Center
NDEP	National Defense Education Program
NDIA	National Defense Industrial Association
NDRI	National Defense Research Institute
NDSEG	National Defense Science and Engineering Graduate Program
NICOP	Naval International Cooperative Opportunities in S&T
NLCCG	Navy Laboratory/Center Coordinating Group
NLCOC	Navy Laboratory/Center Oversight Council
NPS	Naval Postgraduate School
NRAC	Naval Research Advisory Committee
NRC	National Research Council
NRL	Naval Research Laboratory
NSF	National Science Foundation
NSPS	National Security Personnel System
NSTAR	Naval Research-Science & Technology for America
NSTC	National Science and Technology Council
NSWC	Naval Surface Warfare Center
NUWC	Naval Undersea Warfare Center
OECD	Organization for Economic Cooperation and Development

OMB	Office of Management and Budget
ONR	Office of Naval Research
OSD	Office of the Secretary of Defense
OTA	Office of Technology Assessment
PAS	Performance Appraisal System
PBX	Plastic Bonded Explosive
PCAST	President's Council of Advisors on Science and Technology
QDR	Quadrennial Defense Review
R&D	Research and Development
RDT&E	Research, Development, Test and Evaluation
RMs	Reactive Materials
ROI	Return on Investment
SAB	(Air Force) Scientific Advisory Board
S&E	Science and Engineering
S&Es	Scientists and Engineers
S&T	Science and Technology
SCI	Science Citation Index
SECDEF	Secretary of Defense
SECNAV	Secretary of the Navy
SES	Senior Executive Service
SMART	Science, Mathematics, and Research for Transformation
SPAWAR	Space and Naval Warfare Systems Command
SSC	SPAWAR Systems Center
STEM	Science, Technology, Engineering, and Mathematics
STEP	(National Academies' Board on) Science, Technology, and Economic Policy
SYSCOM	Systems Command
TD	Technical Director
TOA	Total Obligation Authority
UARC	University Affiliated Research Center
USD (AT&L)	Undersecretary of Defense for Advanced Technology and Logistics
USNA	United States Naval Academy
VCNO	Vice Chief of Naval Operations
VDP	Virginia Demonstration Project
VSIP	Voluntary Separation in Pay
WCF	Working Capital Fund

Contents

List of Figures

List of Tables

Chapter 1

Introduction

1.1 The New National Security Paradigm

The collapse of the Soviet Union left the United States as the sole world superpower. Even so, the years following the end of the Cold War have witnessed a rapidly shifting global security environment, one now filled with threats such as weapons of mass destruction in the hands of rogue states and non-state players such as al-Qaeda.

Congress authorized the U.S. Commission on National Security for the 21st Century based on a conviction that the entire range of national security policies and processes required examination in light of these new circumstances. In its September 1999 Phase I report, the Commission issued a prescient warning: *"we should expect conflicts in which our adversaries ... will resort to forms and levels of violence shocking to our sensibilities.*[1] Indeed, within two years of the Commission's warning, the tragic events of September 11, 2001 graphically demonstrated the need for a new national security strategy. These events also gave urgent impetus to a sweeping new defense transformation effort aimed at revolutionizing the way the DOD buys weapons, fights wars, and manages its military and civilian personnel.

The new national security paradigm, focused on fighting a global war on terrorism (GWOT), has several implications. For one, there will likely be less emphasis on acquiring large numbers of major weapons systems and more emphasis on novel technologies that address threats such as "dirty nukes" and improvised explosive devices IEDs. In addition, solutions will come more frequently from innovation gleaned from a globalized S&T base that is growing daily, and many will result from breakthroughs that inter-disciplinary and multi-disciplinary research have enabled. Solutions must be found quickly because threats are capable of evolving quickly, and the costs of failure may be incalculable. Nevertheless, solutions must be cost effective, because DOD's research, development, test and evaluation (RDT&E) resources are limited and will more likely decline rather than grow in the future.

1.2   Paradigm Lost?

From the foregoing, the conclusion is inescapable that new breakthroughs in S&T will be required to prevail in this war on terrorism, and the U.S. must maintain its technological leadership in the world. This in turn requires a supply of world-class scientists and engineers (S&Es) adequate for future national security needs. The Commission mentioned above also connected U.S. national security to a strong national S&T effort underpinned by a well-educated science and engineering workforce, and stated in stark terms:

The harsh fact is that the U.S. need for the highest quality human capital in science, mathematics, and engineering is not being met.... This [situation] is not merely of national pride or international image. It is an issue of the utmost importance to national security.

The issue of whether we have enough or the right kinds of S&Es to maintain our S&T leadership is controversial.[2,3,4] Indeed, there is a vigorous and contentious on-going national policy debate about the subject, and authoritative voices on various sides of the question. They include the National Academies of Science and Engineering, the National Science and Technology Council (NSTC), the President's Council of Advisors on Science and Technology (PCAST), and the Council on Competitiveness to mention only a few. Some, such as RAND's Science and Technology Policy Institute, cite supply and demand data, unemployment statistics, and the lack of wage pressure to argue that we have an overabundance of S&Es.[5] Others put forth a more nuanced picture, suggesting, *"some of the numbers and trends about enrollments and degrees are at odds with the conventional wisdom, whereas others show a cyclical pattern with both slumps and spurts."*[6] A number of these divergent viewpoints were presented in a pan-organizational summit sponsored by the Government-Industry-University Research Roundtable of the National Academies.[7] One conclusion that emerges from this debate is that your answer depends upon the question you ask.

It is indisputable, however, that the DOD and other national security-related entities such as the Department of Energy (DOE) and the Department of Homeland Security depend heavily on the physical sciences and engineering, where the data show declines in both federal funding and production of S&Es. Focusing on aggregate numbers masks these trends. It should be of serious concern to the U.S national security enterprise where there is a growing problem in both the public and private sector in finding qualified technical personnel who have required security clearances or who are eligible to obtain one.

The National Defense Industrial Association (NDIA), one of America's leading defense industry associations, has found an astonishing shortage of key specialized workers, both in the defense industry and in the DOD. A 2004 NDIA survey found that almost 9 percent of all funded S&E positions in the defense and aerospace workforce are unfilled due to a lack of qualified candidates [8] and unlike the situation with respect to the general population of S&Es, this shortage has created significant wage pressure. For example, DOD contracts generally allow for a 3.5 percent annual inflation for salaries, but compensation for aeronautical and electrical engineers with security clearances is soaring by 10 to 15 percent annually. Often, efforts to contain these costs have squeezed out innovation by curtailing company Independent Research and Development (IRAD) investments.

Several workforce and other trends contribute to this dwindling pool of cleared or clearable S&Es. Foremost is a lack of interest among American-born youth, especially women and minorities, in the physical sciences, mathematics, environmental sciences, and engineering, at both the undergraduate and graduate levels. A second cause is a graying federal workforce in which more than half of all workers can retire in five years. Government wide, 60 percent of federal employees are over 45, compared with 31 percent in the private sector. Within the DOD, some 43 percent of all civilian workers will be eligible to retire within five years. Within the National Aeronautics and Space Administration (NASA), S&Es over 60 outnumber those under 30 by 3 to 1, a situation largely the result of forced downsizing in the 1990s. [9] Nationally, over half of all S&E degreed workers are 40 or older. Unless there is an increase in degree production, the pool will dwindle more, as baby-boom generation retirements increase dramatically over the next 20 years. A third cause is a long-term decline in the overall Federal investment in R&D as a percentage of Gross Domestic Product (GDP), especially in the physical sciences and engineering. This situation has worsened as a result of reduced DOD S&T funding throughout the 1990s, reductions that occurred despite the importance of these fields to developing new military capabilities.

1.3 Counting the Costs

The negative consequences of diminishing numbers of cleared or clearable S&Es are many. The decline erodes the expertise of the DOD's Service-operated laboratories and centers. Not only has this has contributed to various, costly problems plaguing major defense acquisition programs,[10] but it hinders the overall ability of the centers to carry out what have been three particularly important roles for them since World War II: *conducting long-term, high-risk research private industry is unwilling to pursue; being a quick responder in times of crisis; and measuring technical competence.*

This latter "yardstick" or "smart buyer" function is particularly critical now, as the DOD has already contracted out much of its development and acquisition work and plans to outsource even more. Ironically, increased outsourcing hinders the Department's capability to be a smart buyer at the same time it increases the importance of that capability. To be a smart buyer, the Service laboratories and centers must themselves be knowledgeable performers of hands-on technical work. Yet since the end of the Cold War they have undergone almost continuous personnel and infrastructure cuts, with overall civilian staff reductions often reaching 40 percent or more.

In spite of the problems resulting from continued outsourcing and reductions—problems well understood and documented since the early 1960s—even more reductions are likely. Many cuts so far have been the result of four rounds of the congressionally-mandated base realignment and closure (BRAC) process, and a fifth round will result in even more reductions. The Government Accountability Office (GAO) found that personnel reductions accounted for more than 80 percent of all past BRAC savings,[11] and with the DOD claiming the current round will achieve savings as great as all four previous rounds combined, the burden of cuts will surely fall on civilian personnel.[12] Moreover, because many of the past reductions targeted overhead and support staff, there are fewer of them to eliminate, so many of the cuts will fall on the technical staffs in the laboratories and centers. Given their experience from previous BRAC rounds (and considering the graying of the work force), many S&Es will likely retire rather than relocate. This could have the unintended consequence of damaging vital expertise areas, including some already at or below critical mass, and in which there is no viable private sector alternative.

Consider the following BRAC 1995 example of what could happen this time around: "The process led to considering closure of the Indian Head Laboratory, an East coast site, to move its workload to a West coast site with a test range. Since most scientists and engineers do not relocate with the work, closing it would have devastated a center of critical expertise. That would have cost lives. Only Indian Head had the ability to develop the thermobaric warhead, sparing U.S. troops the bloody prospect of tunnel-to-tunnel combat in Afghanistan."[13]

1.4   Reconstructing R&D

The loss of internal technical competence in the laboratories and centers has become a major concern to many military and civilian leaders who oversee DOD's S&T enterprise. Among the strongest advocates for rebuilding the Department's technical know-how is Dr. Ronald Sega who, until recently, was the Director of Defense Research and Engineering (DDR&E). Dr. Sega advocated the creation of the Science & Math Advancing Research and Technology (SMART) scholarship program to encourage U.S.-born students to seek degrees in science and engineering (the program is fashioned in part after the highly successful National Defense Education Act of 1958, which was prompted by the launch of Sputnik and subsequent fears that the U.S. was falling behind in the space race). In December 2004, the DDR&E, the Aerospace Industries Association (AIA), the NDIA, and an interagency working group of the NSTC's Subcommittee on National Security R&D, hosted a joint Industry-Academia-Government workshop called "National Security Workforce: Challenges and Solutions." It brought together key stakeholders in national security to address ways to find, excite, attract, recruit, and retain skilled workers, technicians, scientists, engineers, and mathematicians critical to U.S. national security.

The question today is not whether we have enough S&Es for our national security workforce needs: there is a problem. The real question is what we, as a nation should do about it. Subsequent chapters will elaborate these issues, especially those involving the in-house laboratories and centers owned and operated by the DON. They will also propose S&T as a new enterprise, and offer recommendations for addressing the other issues that will be discussed.

The solutions we suggest are certainly not the only ones. The important thing, and the aim of this book, is to stimulate an earnest and urgent dialogue about the problem and potential solutions. The time to address the issues is now, while the trouble signs are clearly present but have not yet developed into a real crisis. There are no quick fixes, and acting later will greatly increase the costs. Put simply, acting now will provide a much greater benefit-to-cost (BtC) on a critical element of the investment in defense.

Chapter 2

National Security and the Science and Engineering Workforce

2.1 The Defense Technology Base: An Essential Public-Private Partnership

The defense technology base consists of three sectors: academia and not-for-profits, industry, and in-house government laboratories and centers. Historically, the DOD has relied on universities primarily for basic and applied research, an arrangement with many benefits. Universities perform the largest share of the overall basic research program, a feature that provides access to some of the world's best minds and newest ideas, and also allows DOD to utilize laboratories and other research facilities it does not have to support unilaterally. Funding universities also involves students in defense work and provides in-house laboratories and centers with one of their most important recruiting tools, thereby giving DOD access to the next generation of technologists, engineers, and managers. Similarly, DOD invests in academia to *"to build long term loyalty and interest in DOD problems. A cadre of the nations' best minds, knowledgeable of military problems and willing to consider research of importance to national security, has time and again proved invaluable both in resolving enigmas and opening new opportunities. Academics also maintain contact with their international colleagues in ways that are different in scope and nature from those of government and industry."*[1]

As the defense technology base developed, the DOD set up agreements with several UARCs and established several Federally-funded research and development centers (FFRDCs). FFRDCs had expertise in such emerging areas as radar, space, satellites, and operations research, and were often operated by major universities, since faculty played a major role in their start-up. The Jet Propulsion Laboratory provided an early model of a university-operated facility, while the RAND Corporation was an early model for many of the non-profit technical centers, the so-called "think tanks." Both extend the capabilities of the DOD in-house effort through their ability to attract top technical and managerial talent to work on national security problems, which they do under broad charters to their DOD sponsors.

Universities or privately organized, not-for-profit corporations operate FFRDCs through long-term contracts with the Federal government. DOD currently sponsors eleven FFRDCs managed by eight parent organizations. Each falls under one of three categories: studies and analysis centers, systems engineering and integration centers, and research and development laboratories.[2]

UARCs came into being when the DON entered into a number of memoranda of understanding with certain universities to establish and host laboratories to support research in important areas. In the 1990s the DON reaffirmed its strategic relationship and commitment to four university laboratories to serve as centers of excellence for critical DON and national defense science, technology, and engineering. These four are the Applied Physics Laboratory at Johns Hopkins University, the Applied Research Laboratory at the Pennsylvania State University, the Applied Physics Laboratory at the University of Washington, and the Applied Research Laboratories at the University of Texas Austin.

The UARCs operate under sole-source, multi-task Naval Sea Systems Command (NAVSEA) delivery order contracts to perform work primarily for Navy task sponsors. However, they may also conduct research for the DOD and other government agencies under the NAVSEA umbrella, for programs conducted jointly with the Navy or which have Navy relevance. The special nature of their contractual relationship to the DON allows the UARCs to provide independent technical evaluation to, and serve as technical direction agents for, their DOD sponsors. In addition to their technical work, "*they share their parent universities' public service education and research objectives, and couple their national security interests to their academic skills and resources.*"[3] UARCs are particularly important because they help train new S&Es skilled in areas of importance to the DON.

The other crucial partners in the defense technology base are companies from both the defense and commercial industrial sectors. DOD's in-house laboratories and centers are highly mission-oriented and generally concerned with the entire life cycle of weapons or warfare systems, but only industry manufactures, on a large-scale basis, the products that are the ultimate objective of the development and acquisition process. And while intellectual value is available from a wide range of sources, industry is best equipped to field hardware and support it long term. In the past, industry has provided a number of comparative advantages to DOD. Examples include:

- Integrated technology, systems engineering, and manufacturing expertise, leading to faster fielding of improved capabilities
- A wide range of technological, design, engineering, and manufacturing skills (developed for civilian as well as military products), expanding the range of ideas and options for solving military problems
- Non-governmental international connections, increasing the availability of technologies and reducing the possibility of technological surprise
- Flexible access to top talent
- An ability to project needs. [4]

The last key element of the defense technology base, but by no means the least, is the community of in-house laboratories and centers operated by the military Services. They have a rich history, the roots of some stretching back more than 150 years. Historically, the Navy early on understood the importance of S&E in the conduct of war. It was also among the first to recognize that *"the nature of scientists and 'big science' requires institutional environments to foster creativity and support formulation of ideas and discovery."* [5] Accordingly, early on it began establishing a community of engineering centers, test stations, proving grounds, weapons laboratories, and similar facilities to cultivate these creative environments.

The importance of DOD's in-house technical community has repeatedly been demonstrated and recognized. For example, in October 1961, at the height of the Cold War, Secretary of Defense Robert McNamara declared, *"in-house laboratories shall be used as the primary means of carrying out Defense Department Research and Development Programs."* [6] Some 15 years later, John Allen, Deputy Director of Defense Research and Engineering, succinctly commented, *"No way has been found to preserve the combination of current technical expertise and long-term corporate memory other than setting up an organization wherein individuals can maintain a lasting and close association with their Service while staying involved in technology; in short, an in-house laboratory."* [7]

A study by the White House's Office of Science and Technology Policy, in response to a directive by President Clinton, articulates the laboratories' principal job and value: *"the fundamental responsibility for DOD laboratories is to conduct science and technology programs in* support of national security.... DOD laboratories are best able to translate between technological opportunities and the warfighters' needs, integrate technologies across life cycles and generations of equipment, respond rapidly to DOD needs, provide special facilities, and offer the necessary technical support to the services to make them smart buyers and users of technology."[8]

2.2 The DON Laboratory/Center Community

Today, a community of geographically dispersed warfare and systems centers, along with the NRL, provides most of the internal technical competence to support DON efforts to develop, acquire, and support weapons and weapons systems for the Navy and Marine Corps. This community includes the Naval Surface Warfare Center (NSWC), Naval Undersea Warfare Center (NUWC), Naval Air Warfare Center (NAWC), and the Space and Naval Warfare System Command (SPAWAR) Systems Centers (SSCs) in San Diego, California and Charleston, South Carolina.

This community emerged from the 1991 BRAC round when, on January 2, 1992, the DON formally established several warfare-oriented centers and a streamlined corporate laboratory (NRL) through realigning and/or closing 36 existing R&D, test and evaluation, and engineering centers. When they created this technical community, planners envisioned each warfare/system center would embody within its respective area all the in-house capabilities necessary to support naval systems throughout their life cycle—from S&T all the way through to in-service engineering of deployed systems.[9]

Having a unique mission in a specific warfare or programmatic area, it was decided to assign each center to the systems command with which its mission most closely aligned. The NAWC was therefore organizationally aligned with the Naval Air Systems Command (NAVAIR), NSWC and NUWC with NAVSEA, and the SSCs with SPAWAR. The NRL reports through the Chief of Naval Research (CNR) to the Office of the Secretary of the Navy (SECNAV). Both the centers themselves and their missions are products of a long and complex evolution resulting from changes in the defense environment throughout and since the end of the Cold War.[10, 11, 12, 13]

The SECNAV established a two-tiered group mechanism to oversee and coordinate this community, Figure 2.1.[14] First, the Navy Laboratory/Center Coordinating Group (NLCCG) consisted of the military commanders and civilian directors of the warfare/systems centers and the NRL.[15] Second, members of the Navy Laboratory/Center Oversight Council (NLCOC), chaired by the ASN (RDA), included the

Vice Chief of Naval Operations, the Assistant Commandant of the Marine Corps, the commanders of the naval systems commands, and other senior DON representatives. Its main job was to provide broad oversight of the RDT&E, in-service engineering, and fleet support efforts of the NLCCG. In November 2002, the NLCCG membership was expanded to include the ASN (RDA) and the commanding general and TD of the Marine Corps Warfighting Laboratory which reports to the Marine Corps Combat Development Command (MCCDC).[16]

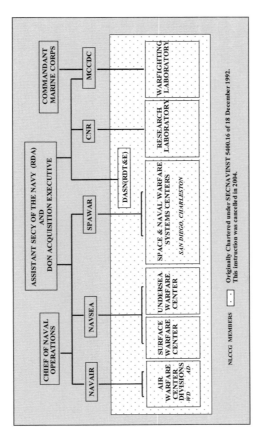

Figure 2.1: NLCCG Organizational Relationships

As previously noted, the warfare/systems centers carry out a broad spectrum of technical work that ranges from basic research through full-scale development of weapons and weapon systems and their in-service support. In doing so, these centers work to ensure: the warfighter gets rapid, direct technical support; the warfighter's needs drive technology investments; a corporate memory exists as a resource (stewardship role); unique facilities are maintained (national investment role); there are unique system engineering capabilities that span multiple warfare areas (interoperability role); and there is a technical broker unaffected by profit motive (yardstick/smart buyer role) and who can be a "supplier-of-last-resort if necessary."

While the warfare/systems centers add value across the entire spectrum of the DON's development, acquisition, and support process, they are in a position to make a unique contribution in that they are best suited to connect the fruits of S&T effort conducted both in-house and in the private sector to create future military capability for the Fleet.

The DOD's S&T program consists of three Budget Activities: Basic Research (BA 1), Applied Research (BA 2), and Advanced Technology Development or ATD (BA 3). Even though S&T funding is only about six percent of the centers' total business base, it still represented $843 million in FY 2004 dollars. About a quarter of the centers' total business base comes from the defense RDT&E appropriation, Figure 2.2.[17]

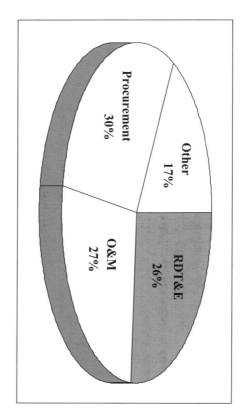

Figure 2.2: FY 04 Center Business Base by Appropriation ($13,982 million) (Source: NLCCG Database)

The DON uses the term "Discovery and Invention" (D&I) to characterize Budget Activities 1 and 2. This effort includes scientific study and experimentation in the physical, engineering, environmental, and life sciences related to long-term naval needs, and efforts (short of major developments) to solve specific naval problems. DON ATD effort supports work to transition the fruits of D&I from the centers into military application in the field. Not all of S&T funding received by the centers is executed in-house. Some of it is contracted out to the private sector to obtain support for their mission work. In FY 2004 only 39 percent of this funding was expended internally.

Regardless of its size, the centers' S&T funding is crucial to their ability to meet their uniquely-assigned roles and missions. It provides the "seed corn" for both the manpower and ideas that lead to next-generation military capabilities. Put another way, S&T historically is where new technologies and their potential applications are explored, developed, and transitioned. It is also vital in that it supplies much of the funding centers use to attract and retain top-quality S&Es, particularly those with Ph.D.s.

The DON's Shrinking S&E Workforce

In spite of their importance, the warfare/system centers have been buffeted by several rounds of consolidation, closure, realignment, and personnel downsizing, as many in DOD believe the private sector should do work once considered inherently governmental. Consider, for example, the 1997 testimony before the House Armed Services Committee of Dov Zakheim, until recently the Defense Department's Comptroller: *"the DOD should be more ruthless about cutting defense laboratories. There is little these laboratories offer that the private sector can't match. While some capabilities are unique to the [DOD], these are far fewer than their proponents will admit, and many hark back to technologies that have long since been bypassed in the private sector...the need for a large defense laboratory structure is simply indefensible....."*[18] John White, Deputy Secretary of Defense in the Clinton administration, and John Deutch, who served as Clinton's Deputy Secretary of Defense and Under Secretary of Defense for Acquisition and Technology, both asserted that the laboratories *"are widely judged to be incapable."*[19] In addition to being inaccurate, such disparaging remarks have affected the image of the in-house laboratories and centers, adding to their difficulties in attracting and retaining top-flight technical staff.

While many of the personnel cuts following the end of the Cold War were inevitable, the way they were implemented was problematic. Most cuts followed the rules of the DOD's industrial-era Civil Service System (CSS), making it all but impossible to target staff reductions and reshape center workforces strategically. Moreover, most of the downsizing focused on *efficiency*—cutting costs. Little attention was paid to the impact of the cuts on *effectiveness*—performing missions. Nor was there much concern over the impact of these reductions on the remaining human capital in the centers. Figure 2.3 compares changes in the NLCCG business base from FY 1991 – FY 2004. After an initial decline, the funding has rebounded vigorously along with the rest of the defense budget, so that by FY 2002, the community had a business base in inflation-adjusted dollars that actually exceeded its FY 1991 total. Over this same period, however, the workforce (civilian plus military) followed a different path. Initially, its decline mirrored the decline in the center's business base as would be expected in an industrial fund setting. However, instead of tracking the turn-around in workload, workforce numbers continued to fall, reaching a reduction of about 43 percent relative to the FY 1991 baseline. In part, this disparity between business and workforce base trends reflects the DOD's increased emphasis on outsourcing. It also reflects a common belief among many defense policy makers that the centers could sustain additional cuts without losing effectiveness.

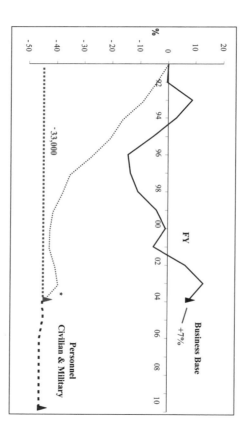

Figure 2.3: NLCCG Community Personnel and Workload Trends (Source: NLCCG Data; OSD Comptroller Green Book of April 2005)

Some warfare/system centers have been reduced in size more than others. Figure 2.4 breaks down the personnel reductions for each of the centers and for NRL over this same timeframe. The NAWC experienced the largest reduction (58 percent) in personnel followed by the NUWC (47 percent). These numbers largely reflect the declines in emphasis in air and submarine warfare relative to other warfare areas following the end of the Cold War.

These reductions have affected support personnel more than the science and engineering (S&E) workforce. The main reason is that many support functions have been contracted out, often in response to congressional, DOD, and Office of Management and Budget (OMB) initiatives to outsource so-called Commercial Activities. Commercial Activities are functions that fall within the purview of OMB Circular A-76.[20] Other reductions have resulted from streamlining initiatives such as business process engineering. In addition, many of the centers' functions—such as public works, base operations, human resources, and finance and accounting—have been turned over to various centrally managed organizations either at the Service or DOD level, under the assumption that such efforts will achieve cost savings. Many NLCCG community activities, for example, are now forced to rely on regionalized human resource offices.

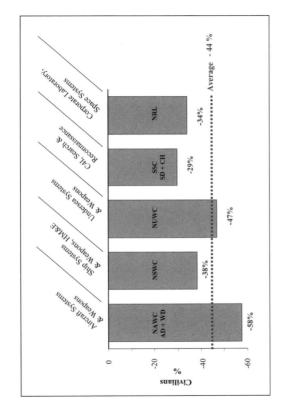

Figure 2.4: Personnel Reductions for Warfare/Systems Centers and NRL (FY 1991 – FY 2009)

As a result of such pressures, between FY 1991 and FY 2004, the overall NLCCG civilian workforce was reduced by 43 percent. Over this same time, the S&E workforce was reduced 20 percent, as shown in Figure 2.5.

In addition to these substantial personnel reductions, intermittent hiring freezes, especially during the 1990s, have exacerbated the situation.[21] By one count, 13 such freezes were imposed on the NLCCG community between September 30, 1990 and March 20, 2002.[22] Some were the result of budget cutting efforts, while others simply mandated an arbitrary workforce reduction, ostensibly to save money. For example, in February 1992 the SECNAV imposed a limited freeze that permitted hiring two people for every five lost. This directive also froze all promotions and/or new appointments into certain high-grade levels (GS/GM 13, 14, 15) until the DON high grade total was again reduced to its September 30, 1991 levels. As another example, in November 1994 the Comptroller of the Navy imposed a 4 percent per year reduction in the number of administrative support personnel as part of the Clinton administration's National Performance Review. This reduction remained in effect through September 30, 1999.

The Internet bubble hiring boom in the private sector also hampered the NLCCG community's recruitment of high-quality S&Es. It was especially difficult to hire at the Ph.D. level in many of the specialties of growing importance, such as nanoscience. In fact, some of the community's best and brightest S&Es left during this period, enticed by higher wages, better benefits, and often by more exciting and challenging work. For example, in a span of just 18 months, DOD lost a key part of its 24-year old ability to perform fiber optics research when industry hired away 26 of NRL's best researchers.[23] At the time, NRL was the Department's only laboratory with this world-class defense capability.

As they have in the past, larger historical events are also driving this trend. Figure 2.6 displays a number of events that produced increased hiring at the centers or their predecessor organizations. Although not shown in this figure, the Soviet's launch of SPUTNIK in 1957 led to a jump in hiring as a result of widespread fear that the U.S. was falling behind in the space and missile race. Other hiring spurts occurred during the Vietnam War era and during the defense build-up led by President Reagan. The downsizing of the 1990s, on the other hand, was the result of the end of the Cold War and the search for a "peace dividend." Recently, President George W. Bush has overseen a major defense build-up spurred in large part by the events of September 11, 2001. Since then, pressure from the growing Federal budget deficit and GWOT costs are

Figure 2.5: NLCCG Community Civilian Workforce Trends (Source: NLCCG Database)

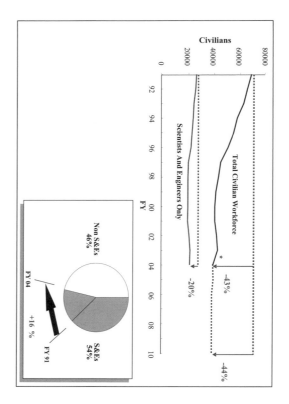

again threatening hiring at the centers, as some of the systems commands turn once more to cost-cutting measures, including limited hiring freezes aimed at shedding personnel.[24]

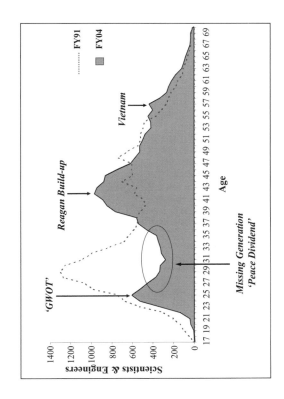

Figure 2.6: NLCCG Community S&E Demographic Profile and Trends (Source: Defense Technical Information Center (DTIC), Defense Manpower Data Center (DMDC))

These and other factors have markedly affected the demographic distribution of S&Es in the NLCCG community and the DON as a whole. This is vividly illustrated in Figure 2.6, which shows demographic profiles of the NLCCG S&E populations in FY 1991 and FY 2004. The number of young S&Es has declined dramatically. Figure 2.7 depicts the age profile of the entire DON civilian community over the period FY 1988 to FY 2000. Again, the aging trends are evident.

Figure 2.8 shows how the various BRACs and mandated reductions have increased the average age of S&Es in the NLCCG community, contributing to a "bow-wave" effect of a generation of experienced technologists moving through the system, without adequate replacements behind the wavefront. The most telling statistic here is that in FY 2004 there were 57 percent fewer S&Es under the age of 31 than in FY 1991. Note also that the average age in FY 1991, 38.2 years, had increased to 42.7 by FY 2001. An increase in hiring in 2001 caused this average age to decrease slightly, to 42.2 years by FY 2003. However, NAVSEA's cost-cutting initiatives in early 2005 caused again restricted S&E hiring and threatened to snuff out the little progress made. As a result, the average age is again on the rise. Perhaps one of the most important trends to note is the steady decline in S&Es in the 30 to 40 age group. There were 6,920 S&Es in this age group in FY 1991, but by FY 2004 this number had dwindled to only 4,801.

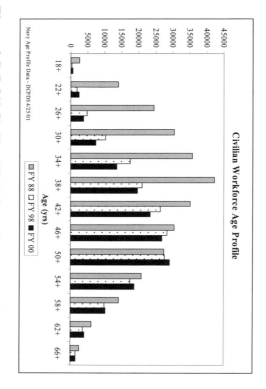

Figure 2.7: DON Civilian Workforce Demographic Profile and Trends (Source: Defense Civilian Personnel Data System (DCPDS) Data)

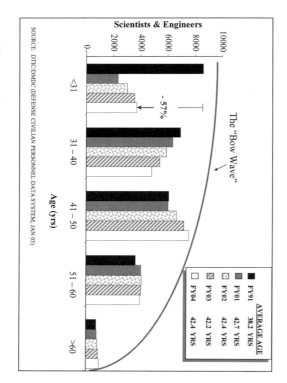

Figure 2.8: NLCCG Community S&E Age Trend (Source: DTIC, DMDC, DCPDS data)

The Negative Consequences of a Shrinking Workforce

These sizeable, recurring, multi-layered, multi-year cuts and the consequent aging of the S&E population have had serious consequences for the defense S&T effort, and threaten more ominous consequences in the long-term. For example, as many analysts are discovering, downsizing alone often carries with it a host of negative effects. These include, to name but a few, adverse effects on employee loyalty, the loss of invaluable corporate memory, organizational instability, and, paradoxically, high costs.[25]

In addition, the S&E aging trends disturbed senior civilian and military leaders enough that, in 2000 and 2001 they prompted several high-level DON assessments of the problem. For example, the SECNAV chartered a task force to study potential DON human resource management systems for the future. The National Academy of Public Administration (NAPA), through its Center for Human Resource Management, helped with the study. The results appeared in "Civilian Workforce 2020: Strategies for Modernizing Human Resource Management in the Department of the Navy," released in August 2000.[26] The study recommended a new human resources management system to replace an aging CSS unsuitable for 21st century needs. The Academy offered other recommendations to allow the DON to acquire a workforce aligned with its future mission requirements, for example: an integrated, strategic planning process; modern information systems support; a restructured management community; and a new strategy for acquiring and retaining talent.

Soon after the release of this NAPA report, the Chief of Naval Operations (CNO) and Vice Chief of Naval Operations (VCNO) initiated a study that also discussed the negative effects of downsizing, reduced hiring, and outsourcing. Responding to a briefing on the DON civilian workforce, the VCNO chartered a "Civilian Manpower and Personnel Management Task Force." In its June 2001 report, the first problem the task force discussed was the effect of workforce aging on demographics and the balance of skill mixes in the DON civilian population. The task force described the problem this way:

This is an outcome of the downsizing of the last decade and the methods we used to achieve that downsizing. Years of [reductions] have produced a force that relies upon experienced workers. We no longer have the on board strength in the younger age groups to naturally replace workers as they retire or as they simply leave the Navy for other work. The civilian workforce is not recruited or sustained in a fashion similar to the military. As the civilian workforce was reduced...fewer young workers were hired.... Voluntary reductions were stimulated through employee buy-outs and incentives for regular and early retirement. Navy met its civilian workforce reductions – as required in BRAC and the past Quadrennial Defense Review (QDR), but we no longer possess the number of younger workers to inherently replace older workers.... This is true both in government and industry for technical, engineering, management and the industrial workforce components. Attrition, in addition to retirement, is a significant issue across the age groups. As an example, in FY-00, over 5,200 retired, but over 7,700 just left Navy, and we hired less than 7,500 people. This net loss in our civilian workforce cannot be permitted to continue.... [It] will decimate our capability....

...we must focus both on attrition as well as retirement, and we must adopt an integrated outsourcing strategy. We need broader adoption of methods to increase both our hiring and retention. An increasing proportion of our workforce has the opportunity to retire in the next decade. We must prepare for an orderly transfer of knowledge. We must reinvigorate our hiring of both new graduate employees as well as more experienced people if we are to sustain the civilian workforce at levels necessary to meet the requirements set for them. We must provide opportunities for professional growth for our workforce; we need to provide for interesting work in order to retain the best and brightest.

Despite the alarms these and other high-level studies sounded, few of the recommendations were implemented, and then 9/11 and its aftermath diverted the attention of senior civilian and military leaders. The problems have not gone away, however, either for the NLCCG community or the DON. Indeed, current trends suggest they will continue to worsen unless significant, sustained steps are taken to reverse them.

The Nation's Shrinking S&E Workforce

The demographic trends in the warfare/systems centers are in fact indicative of a much wider problem. Comparable troubles are affecting all the Services, other Federal departments and agencies that operate laboratory systems (such as DOE and NASA), and a growing number of defense and aerospace companies in the private sector.

Sean O'Keefe, until recently the NASA Administrator, discussed these trends in his testimony before the House Science Committee in July 2002.[27] Within NASA's S&E workforce, the over-60 population outnumbered the under-30 population by almost 3 to 1. The age contrast was even more dramatic at some of NASA's field centers. For example, at the Marshall Space Flight Center only 62 engineers out of a 3,000-person workforce were younger than 30. Similar proportions prevailed at the Glenn and Langley Research Centers, where the corresponding ratios were 5:1 and 7:1 respectively. O'Keefe noted that by contrast, "*in 1993 the under-30 S&E workforce was nearly double the number of over-60 workers. This is an alarming trend that demands our immediate attention with decisive action if we are to preserve NASA's aeronautics and space capabilities.*"

Figure 2.9, taken from O'Keefe's testimony, shows how the demographic profile for NASA S&Es has changed from 1993 to 2002. These data clearly demonstrate that the space agency has significantly fewer young S&Es than ten years ago.

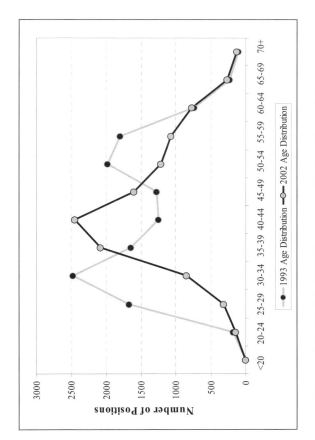

Figure 2.9: NASA S&E Demographic Profile and Trends (Source: Congressional Testimony by Sean O'Keefe)

Demographic data gleaned from the DOE laboratory community tell a similar story, as that Department has begun to experience hiring, recruitment, and retention problems within its laboratory system. This community includes the so-called national laboratories, which have key missions in basic science, national security, energy resources, and environmental quality. These laboratories have long been regarded as among the most eminent in the world, and have also been a critical factor in helping the U.S. maintain its worldwide leadership in generating scientific knowledge and discovery.

Recently, however, the DOE laboratories, like those in NASA and DOD, have experienced critical challenges regarding their S&E workforces. The Department established a task force to recommend actions for consideration by management, an effort involving almost all of the DOE laboratories.[28] Though there are differences among individual and among groups of laboratories (for example national security versus science laboratories), the task force focused on the high degree of commonality across the laboratories on workforce issues.

In December 2000, the task force published its findings in a draft white paper, highlighting several factors making it more difficult to maintain world-class technical staffs and hire new S&Es with skills critical to the future.[29] For one thing, the paper identified a perception of the erosion of exciting scientific and engineering work in the laboratories. For another, it cited funding instability caused by on-going fluctuations in the DOE budget as an impediment to strategic workforce planning. In effect, these fluctuations resulted in erratic hiring patterns. Moreover, the panel expressed concern over the erosion of flexibility resulting from increased restrictions in personnel management appendices to the laboratory management contracts.

Another overarching problem the task force identified—one hurting almost all members of the defense technology base involved with classified projects—was the S&E pipeline. There are simply fewer high technology graduates with eligibility (primarily clearances) to work on national security problems. This security clearance impediment is a recurring theme throughout this book.

Also worrying to the task force was an atmosphere in the laboratories that could discourage foreign nationals and even Asian Americans from working there. The task force recommended that, "*a cautious and rational security policy must be crafted to allow the Labs to tap into the outstanding scientific and engineering talent pool in this country, including foreign nationals.*" It further bemoaned what it called a worsening atmosphere with a drift towards compliance-based management.

The task force's demographic analysis also found an S&E population aging much like that in the DON. Even after accounting for differences in degree levels, the DOE laboratories had a significantly smaller proportion of S&Es under the age of 40 than the U.S. norm (26 percent versus 40 percent), with the largest percentage of its S&Es between 41 and 50 years old, Figure 2.10.

The problems the DOE task force identified have been addressed in numerous studies of the DOD laboratories and centers. One example is a recent Naval Research Advisory Committee (NRAC) report, chartered by the DDR&E, called "Science and Technology Community in Crisis."[30] It notes that a world-class research institution—one with an outstanding technical staff, important and challenging work, state of the art facilities and equipment, and visionary leadership—cannot exist without the ability to attract, retain, and reward high-level talent. The primary culprit preventing the recruitment of such talent is the outdated CSS, as it cannot be adapted to specific needs. The report's major recommendation calls for replacing the personnel system with one tailored to a research institution, and points out that Congress has already given the DOD the tools to set up such a system.[31]

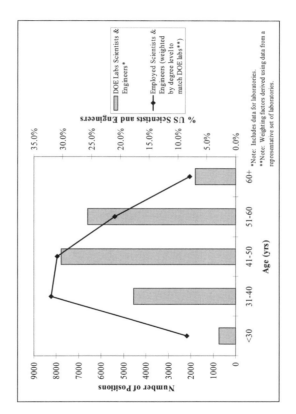

Figure 2.10: S&E Demographic Profile in DOE Laboratory Community (Source: DOE White Paper)

The demographic trends that imperil DOD's laboratories and centers also threaten a growing number of private sector defense and aerospace companies, and this too has serious implications for U.S. economic and national security. Responding to this problem, the Air Force asked the National Research Council (NRC) of the National Academies to "*provide a report that addresses the effects of U.S. defense industrial* base shrinkage and the aerospace industry's ability to continue to attract and maintain requisite aerospace engineering talent...to produce cutting-edge military products" the DOD needs.[32]

The report emphasizes how current trends threaten the inter-generational transfer of specialized, crucial technical knowledge and skills, a point that also directly applies to the in-house laboratories and centers:

The change in the age-experience composition of the work force occasioned by the decrease in defense spending raises serious questions about mentoring and the generational passing on of knowledge in the industry. One immediate effect is that older employees who qualify for early retirement may elect to retire because they see fewer opportunities ahead for interesting work. Meanwhile, younger engineers may leave for what they perceive to be better opportunities...elsewhere. In addition, the short-term result would be that the work force is predominantly middle-aged. As time passes with no new significant hiring, the work force will become disproportionately composed of older, more experienced employees. This has occurred in the aerospace industry and in government over the past 15 years, during which time the number of engineers aged 25 to 34 has fallen from 27 to 17 percent of the work force. In the space sector, only 7 percent of the engineers are under age 30. At Lockheed Martin Aeronautical Company, for example, new hires, which started to decline in the early 1980s, dropped to almost zero in the 1990s. If this trend continues, as experienced workers age and retire, their knowledge and expertise will be lost.... If the need is there, which seems extremely likely, the gap must be filled by young and inexperienced people.

Many companies are now planning formally arranged mentoring procedures, as well as more training programs for new workers. When new engineers are hired, mentoring and teaming arrangements have to be carefully planned to capture the experience of those about to retire. Nevertheless, experience is lost, as is efficiency, when work tasks involve significant learning curves."

Various other reports have discussed these and similar problems in the aerospace industry. One commentary is in the report of the congressionally-established Commission on the Future of the United States Aerospace Industry, issued in November 2002.[33] Even more recently the AIA has proposed steps to address the difficulties, as have others.[34]

As noted in the Introduction, the NDIA, an association of defense companies, has joined the chorus of those alarmed over the dwindling supply of cleared or clearable S&Es who possess critical skills. In a 2004 survey of defense companies, they found thousands of unfilled S&E positions for U.S. citizens, with the situation getting worse.[35] The most acute shortages were in engineering disciplines, principally aerospace, software, electrical, and mechanical. However, shortages were also identified in mathematics, physics, chemistry, and materials. The study does not describe a "*shortage environment*," because data show many areas where the supply of S&Es exceeds demand. Rather, the "*requirement for security clearances [is] greatly restricting the supply side for DOD/government*." The authors worried that the conjunction of three trends—the retirement of the post-Sputnik generation, the decline in clearance-eligible S&Es, and the diminishing U.S. technological dominance due to globalization of R&D—could lead to a "perfect storm."[36]

In sum, the foregoing discussion clearly demonstrates that finding qualified technical personnel is a problem of growing concern to the U.S. national security enterprise. Exacerbating the trouble is the shortage of people who have required security clearances or are eligible to obtain one: eligibility for most national security jobs requires U.S. citizenship. What the DOD can do about the situation is a subject reserved for a later chapter.

2.3   Declining Center Science and Technology Resources

Unlike the NLCCG community's civilian workforce, other components of the DON enterprise have experienced a more positive reversal of fortune over the last few years. We have already noted that the overall business base of the NLCCG declined from FY 1991 to FY 1996. Thereafter, it grew markedly, reflecting increases in the overall DON budget measured in terms of TOA. Figure 2.11 illustrates that DON TOA also declined significantly from FY 1991 to FY 1997. It then began to increase and by FY 2003 had returned to its inflation-adjusted FY 1991 level. In fact, the latest budget projections suggest TOA in FY 2010 will be about 2 percent greater than in FY 1991. Figure 2.11 also shows that while workforce levels in the defense industrial sector track these budget changes, those in NRL and in the warfare/systems centers do not. This is partly a result of factors already discussed, and partly a result of mandated workforce reductions imposed on the centers.

Nor have the trends in the centers' S&T funding—one of their most important resources—tracked the overall budget increases. While S&T funding represents only about 6 percent of the centers' total business base, it is essential if they are to carry out their assigned roles and missions. That is, it supports breakthrough research and also funds the transitioning of that research into next-generation military capabilities. Just as important, S&T funding provides the challenging work needed to attract and develop new research talent, a point emphasized in the DOE white paper mentioned above: *"Exciting work and a promising future are required to attract and retain employees with critical new skills.... basic sciences and exploratory research and development...keep the Labs at the forefront of cutting-edge technology and thereby reenergize their workforce."*

Table 2.1 shows how the DON Basic Research, Applied Research, and ATD dollars have changed from FY 1992 to FY 2004. Overall, S&T dollars increased, in significant measure the result of additions by the Congress during its mark-up process. However, as a percentage of TOA, S&T still remains well below the overall DOD goal—three percent of TOA. Most of the growth in S&T has occurred in the ATD account, which grew by almost 260 percent over the period FY 1992 to FY 2004.

Regrettably, center S&T funding has not even come close to keeping pace. For example, in FY 1992 the centers captured 45 percent of the DON S&T dollars. However, by FY 2004, their share had fallen to just 22 percent.

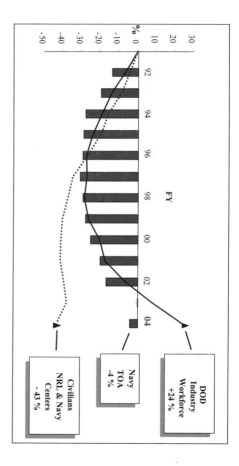

Figure 2.11: Civilian and Defense Industry Workforces and DON Budget (Source: NLCCG Database; OSD Comptroller Green Book of April 2005)

Fiscal Year	Basic Research	Applied Research	Advanced Technology Development	Science and Technology	Science and Technology Percent of DON TOA
1992	476	593	289	1,359	1.24
2004	468	678	1,036	2,182	1.79

Table 2.1: DON S&T Funding Trends in Millions of Constant FY 2004 Dollars

The potentially ominous consequences of such a drop are particularly apparent in the funding trends for D&I. In FY 1992, D&I represented 79 percent of DON S&T. However, by FY 2004, it represented only 53 percent, a decline of some 33 percent. This means the centers did less S&T focused on developing new knowledge related to long-term military needs, and more on transitioning already existing technology into application. In sum, D&I represents the "seed corn" for future capabilities, and current trends are increasingly nibbling away at it. Robert Frosch, a former Assistant Secretary of the Navy for research and development, described this situation as unsustainable in the long run, akin to a farmer who wishes only to harvest and not to sow.[37]

Another feature of the centers' S&T funding has to do work carried out in-house versus that contracted out to the private sector. Of the approximately $491 million in DON S&T they received in FY 2004, the centers expended some 56 percent on work done in-house and contracted out the other 44 percent. It should be noted that the percentages vary across the three S&T budget activities, as shown in Table 2.2.

Center Budget Activity	($ Millions of FY 2004 Dollars)			Percent Performed In-House
	In-House	Out-of-House	Total	
Basic Research	20.3	4.3	24.6	83
Applied Research	105.1	62.3	167.4	63
Advanced Technology Development	147.9	150.7	298.6	50
Total S&T	273.3	217.3	490.6	56

Table 2.2: Who Performs the Centers' DON Science and Technology Work?

2.4 Discretionary S&T Funding: The Quintessential Element of Workforce Revitalization

Yet another threat to DON's overall S&T program is the drop in discretionary S&T funding. Discretionary funds are allocations laboratory and center directors use, with some constraints, as they see fit. These funds give directors more flexibility in pursuing promising projects, and therefore more opportunities to attract high-quality personnel. The steady decline in such funding over the last dozen years is further eroding capabilities to carry out work that only defense laboratories and centers can do, and which is critical to national security.

For over 40 years planners and officials have recognized that only the defense laboratories and centers can perform certain S&T work essential to national defense. In fact, DOD has been alarmed over the deterioration of such capabilities in the past, and acted to rectify the problem. For example, in April 1962, David Bell, director of the Bureau of the Budget (now the OMB), submitted a report to President Kennedy on contracting by the Federal government for research and development.[38] The study had been prompted by concerns that excessive contracting out had blurred the lines between those technical functions that are public and those that are inherently governmental. Bell argued:

The decisions which seem to us to be essential to be taken by government officials, rather than being contracted out to private bodies of any kind, are...what work is to be done, what objectives are to be set for the work, what time period and what costs are to be associated with the work, what results expected are to be, and...the responsibilities for knowing whether the work has gone as it was supposed to go, and if it has not, what went wrong and why, and how it can be corrected on subsequent occasions.

In essence, Bell was describing what has since been called the "smart buyer" or "yardstick" capability. In the DOD, most of the competence needed to perform this role resides in the technical staffs of the laboratories and centers the three Services operate. In this regard, the Bell Report stressed that the government should *"have on its staff exceptionally strong and able executives, scientists, and engineers fully qualified to weigh the views and advice of technical specialists."* It bemoaned the *"serious trend toward eroding the competence of the government's research and development establishment—in part owing to the keen competition for scarce talent which has come from government's contractors."* Bell concluded that it is *"highly important to improve this situation by sharply improving the working environment in government, in order to attract and hold first-class scientists and technicians.''*[39]

A later report by the DOD Inspector General characterized the concerns that prompted the Bell Study as follows:[40]

The Bureau of the Budget concluded that the substantial increase in contracting out for research had seriously impaired the Government's in-house ability to execute research and development work. The DOD, in particular, had come dangerously close to permitting contractor employees to perform functions that were the responsibility of Government officials. In addition, the Government's in-house ability to supervise and evaluate research and development efforts that were contracted out had been seriously impaired. [This] was caused by the best of the Government's research scientists, engineers and administrators being recruited by private industry as a result of higher salaries, significant and challenging work, and better working conditions.

The Bureau of the Budget recommended that laboratory directors be provided a discretionary allotment of research funds to reverse this trend. The laboratory directors were to use the discretionary funds to strengthen the internal competence of individual laboratories by assigning research projects that were significant and challenging enough to attract and hold competent personnel. Laboratory personnel were also to be given greater participation in program determination and in providing technical advice. In summary, laboratory directors were to be given discretionary funds and the authority to make decisions relating to research projects, personnel, funds, and other resources. The [ILIR] Program was initiated Defense-wide in FY 1963, and the DOD Manual 7110-1-M, "Budget Guidance Manual," provides the guidance on the use of Program funds.

There was a general understanding at the time that the laboratory directors needed research funds to provide the kind of in-house work that would attract and retain high quality staff. In fact, while the Bell Study was in process, the underlying policy for establishing both the discretionary In-House Laboratory Independent Research (ILIR) program and a complementary BA 2 discretionary Independent Exploratory Development (IED) program, was laid out in a memorandum from Secretary of Defense McNamara. In this same memorandum McNamara also declared, "*in-house laboratories shall be used as the primary means of carrying out Defense Department Research and Development Programs.*"[41]

As mentioned, direction for applying ILIR and IED program dollars was included in the DOD Budget Guidance Manual, which directed that funds will be provided:[42]

…to support, in addition to regularly assigned programs, work judged by the Directors of these RDT&E laboratories to be important or promising in accomplishment of assigned missions…it is intended that Laboratory Directors be given the widest latitude…since the purpose is to enable the Director to perform innovative, promising work without the procedure of formal and prior approval…

Subsequent DOD policy interpreted this budget guidance in more detail:[43]

- Each DOD Component that operates a research and development Laboratory or Center, shall support an ILIR program. In addition, each DOD Component may support an IED program.

- The technical director and/or commanding officer (TD/CO) of each participating laboratory or center shall be provided with ILIR and IED funds to initiate and support efforts judged to be important or promising in the accomplishment of missions assigned to that laboratory and/or center.

- Each…TD/CO shall be given wide latitude in the use of ILIR and IED funds subject to the approval of overall funding levels, to enable performance of innovative, timely, and promising work without requiring formal and prior approval that might delay normal funding authorization.

- ILIR funds shall be used to support basic research (6.1) efforts and IED funds (if available) shall be used to support exploratory development (6.2) efforts. These programs shall be used to support exploratory development in support of laboratory missions, and enhancement of factors that contribute to recruitment and retention of outstanding scientists and engineers.

For decades, the significance of discretionary funding was widely accepted. For example, a 1983 blue ribbon report of the White House Science Council, a study chaired by David Packard, recommended that "*at least 5 percent, and up to 10 percent, of the annual funding of the*

Federal laboratories should be devoted to programs of independent research and development at the laboratory directors' discretion.[44] In a 1987 study on technology base management, the Defense Science Board (DSB) asserted that, *"a successful laboratory requires discretionary basic research funding for its long term viability."*[45] The DSB recommended that *"at least 5 percent, and up to 10 percent, of the annual funding of Federal laboratories"* should consist of ILIR funds, the same percentage recommended in the Packard study.

The Congress agreed. Worried that the U.S. might be losing some of its technological superiority over the Soviet Union, Congress asked its Office of Technology Assessment (OTA) to assess the health of the defense technology base. In April 1989, the OTA issued a comprehensive report called *Holding the Edge: Maintaining The Defense Technology Base* that addressed, among other things, technology base funding, including that of the ILIR and IED programs.[46] The OTA observed:

The ILIR programs serve a number of important purposes.... Because they are a principal main source of discretionary research funds, the Service ILIR programs help the laboratories maintain an atmosphere of creativity and research excellence, enhance their science and technology base, provide seed money that can lead to new research efforts, and assist the laboratory directors in hiring new PhDs.

Despite the broad agreement on the importance of discretionary funding in general, and ILIR and IED funding in particular, both programs suffered erosion in their funding base beginning in the late 1960s. From FYs 1967 through 1980, ILIR funding at the Navy's RDT&E centers decreased by 59 percent, while IED funding decreased by 74 percent (adjusted for inflation). Although the DOD budget guidance mentioned above specified that ILIR and IED should not constitute more than five percent of a laboratory's funds, only in 1967 did the two programs even approach five percent. In fact, ILIR and IED expenditures, as a percent of the total budgets of the Navy RDT&E centers, declined significantly after that—between 1967 and 1980, these programs dropped from 3.8 percent of total funds to just 1.5 percent.

The cancellation of the IED program in 1993 had a particularly adverse impact on the centers' discretionary funds. In part, an overall reduction that year of the DON S&T account caused the termination. However, it was also the result of a historically demanding defense of the program to Congress. It was always difficult to explain to congressional staff why the IED program involved only after-the-fact review and oversight.[47]

The IED program, like ILIR, was a relatively small account, usually around $25 million a year for the DON. Moreover, when this was divided among the various center sites, the individual allotments were quite modest in proportion to the overall business base of the sites. An example is in Table 2.3, which shows the size of the IED program at NSWC's Dahlgren Division for FY 1989 through the last year of the program.[48] Note that most projects received about $100,000, enough on average to support one scientist or engineer for a year.

Table 2.3: Independent Exploratory Development Funding at NSWC's Dahlgren Division

	FY 1989	FY 1990	FY 1991	FY 1992	FY 1993
Funding (Thousands of Dollars)	2600	2412	2880	1437	1219
Number of Projects	24	22	25	15	13
Average per project (Thousands of Dollars)	108	110	115	96	94

Still, the IED program was especially important as a source of discretionary funds for development of ideas generated in the ILIR and other basic research programs, often scaling up the nature of the basic work to a more realistic physical level.

In sum, despite the relatively modest sizes of both the ILIR and IED programs, they have in the past served important functions, including:

- Providing funding to the centers for basic and applied research in areas important to their missions
- Enabling innovation
- Developing and maintaining a cadre of S&Es capable of tracking and evaluating the rapidly growing global data base of research and new knowledge in order to apply it to problems of naval interest
- Promoting the hiring and development of S&Es

- Encouraging and supporting cooperation with universities, industry, the NRL, and other DON and DOD laboratories.

Both programs also have proven track records of productivity, measured in terms of output metrics such as technical papers published, patents applied for or received, and awards and honors. Metrics for the ILIR program are shown in Table 2.4.

Even more important is the overall impact of these programs over the years in transitioning the results of new discoveries and inventions into weapons and warfare systems.

The cancellation of the IED program, and the general decline in S&T funds (especially discretionary funds), have been a major blow to the technology base of the centers and have contributed to several worrisome trends. One, these changes have contributed to the workforce shortages discussed above, pushing many S&Es who worked on IED efforts into other paths of work. Similarly, lack of discretionary resources undermines laboratory directors' ability to provide the challenging opportunities that attract and retain expert technical staff. From 1995 to 1999, the centers experienced a 15 percent reduction in the number of S&Es with advanced degrees (M.S. and Ph.D.). Figure 2.12 illustrates the decline in Ph.D.s as a percentage of new S&E hires in both the warfare/centers and NRL. For example, among the new hires in FY 1997, 45 percent at NRL were Ph.D.s while only 5.6 percent of those at the centers had doctorates. For the class of FY 2004, these percentages had dropped to 39 percent and 3 percent respectively. While the NRL data show several ups and downs in Ph.D. hiring patterns, the data for the warfare systems centers show a slow but steady decline in the percentage of new S&E hires with doctorates. In fact, from FY 1997–FY 2004 that percentage declined by about 46 percent.

In FY 2004, the NLCCG community's civilian workforce included 20,795 S&Es, but only 1,702 of them had Ph.D. degrees— slightly more than 8 percent. Table 2.5 shows the percent change in the Ph.D. populations at the warfare/system centers and at NRL from FY 1997 to FY 2004. The community as a whole experienced almost a 15 percent decline.

As a result, the overall intellectual capacity of the centers is threatened, a situation reflected in the output of their ILIR programs: they produced 225 published papers and 72 patents/patent applications in FY 1996, but by FY 2003 the corresponding figures had dropped to 77 and 38 respectively.

Output Metric	'93	'94	'95	'96	'97	'98	'99	'00	'01	'02	'03
No. Transitions	--	38	35	41	33	38	44	49	50	32	68
No. Projects	181	197	218	214	206	192	183	195	198	181	151
Published Papers	393	270	259	225	250	261	270	202	203	142	77
Submitted Papers	110	90	82	74	59	17	19	54	109	56	17
Books/Chapters	52	7	20	19	6	8	13	16	8	11	3
Patents/Patent Applications	79	67	69	72	93	98	99	105	94	99	38
Government Reports	76	62	68	56	38	27	20	8	19	10	8
Dissertations	37	10	13	5	3	1	6	5	3	2	2
Presentations	417	259	369	248	248	234	234	229	220	198	76
Awards/Honors	79	52	68	39	41	39	48	19	14	39	15
Funding (Millions of dollars)	15.4	16.8	17	15	13.7	13.1	13	14.2	14.4	15.6	14.0

Table 2.4: ILIR Program Metrics

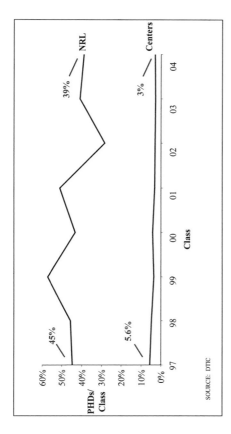

SOURCE: DTIC

Figure 2.12: Ph.D. Hiring Trends at the Centers and NRL (Source: NLCCG Database)

Lab/Center	FY 1997	FY 2004	Percent Change
NAWC	267	216	-19.1
NSWC	460	404	-12.2
NUWC	143	125	-12.6
SSC	201	162	-19.4
Center Total	1,071	907	-15.3
NRL	930	795	-14.5
NLCCG Community Total	2,001	1,702	-14.9

Table 2.5: S&Es with Ph.D. Degrees at the Centers and NRL

Further, such reductions undercut the centers' ability to conduct fundamental research in militarily important emerging areas, including new fields of interdisciplinary research. Examples of areas that could be affected include nanoscience and advanced materials (e.g., biology-based materials, miniature systems, new energetics, advanced electronics), directed energy (e.g., high-temperature lasers, high-power microwaves, pulsed power), and advanced power (e.g., batteries, energy storage, generation and handling of electric power).

The loss of staff with advanced degrees, left unchecked, will also reduce the centers' ability to judge the products the DON receives from the private sector. In other words, their yardstick role in emerging areas of science, engineering, and mathematics will become more difficult at the very time the need for this internal competence is growing. As has been seen, the centers already outsource a large percentage of their S&T effort to obtain support for the in-house component. To assess that work, they must be able to interface, peer-to-peer, with their private sector colleagues conducting it.

2.5 A Growing Demand for In-House Science and Technology Capability

Although S&T resources, especially discretionary resources, going to the warfare/systems centers are declining, there are several reasons to believe the demand for center capability will grow in the future. These reasons are mentioned briefly below, and then each is discussed in detail.

One reason to believe this demand for S&T effort will increase is that the continued growth in outsourcing will eventually require a commensurate strengthening of the centers' yardstick competence. The Federal Government must have objective technical advice about the quality, military relevance, and overall worth of the contracted work. This can be obtained only through sources insulated from pressures to profit. Otherwise, it cannot be an intelligent consumer of private sector products. Indeed, the absence of such advice could waste precious defense resources or, worse yet, undermine national security.

A second factor indicating a future increased demand for center S&T capability is the growing reluctance of defense companies to invest their own resources in long-term, high-risk research. This is part of a general trend that has affected most commercial technology companies, and it could mean the DOD will have to look elsewhere, including to its own Service laboratories and centers, for many of the innovations needed in the global war on terrorism and other conflicts. A recent article in the press addressed this issue, noting, *"It is not unusual to hear defense officials complain that contractors are too focused on their financial bottom lines, rather than on the quality of their new products and the needs of the customer. They also blame the industry's rapid consolidation into a handful of conglomerates for a perceived decline in technical innovation."*[49]

A third factor involves the remolding of the industrial base by the on-going process of defense transformation Secretary Rumsfeld has begun. The point here is that the effort could result in less DOD investment in big-ticket weapon systems and platforms, a development that would jeopardize the funding base of today's major defense contractors. This could, in turn, pressure these companies to further reduce all investments, including those in R&D. Even more dramatically, it could force a number of them out of business. The implication in either case is that other sources may have to take up the slack, which could shift work back to the laboratories and centers.

A fourth factor that could interest defense policy makers in increased in-house performance of S&T is an inability, or in some cases reluctance, in academia to pursue certain kinds of defense research. In part, this is due to causes stemming from the events of September 11, 2001, such as increased scrutiny of foreign students at both the graduate and undergraduate levels and reductions in student visas.

A final factor that militates in favor of increased in-house performance of S&T is the globalization of the technology base. Most likely this will intensify demands on the centers' capability to track, assess, and apply this rapidly growing base of new research and knowledge.

Each of these factors is discussed in more detail in the following sections. Whether or to what extent any or all of them will have the impact suggested is yet to be determined. Nevertheless, they seem to increase the likelihood that the DON warfare/system centers will need more, not less, in-house S&T capability in the months and years ahead.

2.6   Increased Outsourcing Calls for a Bigger Yardstick

Over the last 50-plus years, there have been innumerable authoritative statements regarding the proper roles of DOD's in-house laboratories and centers; these statements show a high level of consensus with respect to several of those roles. Table 2.6 summarizes a few examples, but there are many others, most containing variations of the same or similar themes. For example, almost all studies contend that the laboratories and centers exist to:

- Enable the DOD to be a "smart buyer" in the systems acquisition process
- Provide technological expertise in areas of limited interest to the private sector
- Provide an immediate response in time of crisis (wartime, for example)
- Maintain a corporate research and development memory
- Maintain and provide specialized equipment and facilities impractical for the private sector to provide for itself

Jim Colvard, a prominent former Navy laboratory TD, Deputy Chief of Naval Material, and Deputy Director of the Office of Personnel Management during the Reagan administration, has written extensively about in-house laboratories, often about their smart buyer role. He notes that while many in the DOD today believe you can go directly to industry with a problem, it is not that simple—a fact attested to by the billions of dollars DOD has poured into contractor claims and get-well programs over the years.[50]

Colvard describes the absolute necessity of the smart buyer role in these practical terms: "The Navy can never contract out its ability to understand military problems in technical terms, know who has the potential to solve those problems, and be able to verify a correct solution technically when it is presented." He also argues that what provides these capabilities is the interfacing of the laboratories' and industry's technical infrastructures.

"Without that technical capability, the Navy finds itself in the position of having an administrative interface—short on technical understanding—dealing with industry."

Another aspect of this relationship Colvard mentions is that military preparedness is a continuous function, and the rapid response and corporate memory roles of the laboratories are critical adjuncts to their smart buyer role. *"The retained intellectual residuals from investment in the Navy's science and technology infrastructure are available on demand to the Navy.   Knowledge and experience gained through a contract operation may well be lost when the contract ends or goes to another contractor."*

The DOD's need for the kind of technical competence Colvard describes is growing daily because of the large amount of technical work contracted out to the private sector. As the Department moves to outsource even more of its work, it will increasingly need an internal yardstick to assess whether the product it is getting is what it is advertised to be, and not just a cheaper—not to mention useless— substitute.

Paradoxically, DOD's increased outsourcing of technical work makes it more difficult for the centers to assess technical competence, because to do so, they must be knowledgeable performers of hands-on technical work themselves. The more work contracted out, the greater the importance of a basic level of in-house technical proficiency. In the increasingly complex world of S&T, this means having S&Es who are themselves experts in fields relevant to the DOD's current and future needs.

National Security and the Science and Engineering Workforce

Table 2.6: Examples of Roles Performed by DOD Laboratories and Centers

White House Report (1979)[51]	Director of Navy Labs Report (1980)[52]	Federal Advisory Commission (1991)[53]	White House Report (1994)[54]
– Yardstick or smart Buyer – Mission-oriented studies, tech analyses and evaluation – Corporate memory – Independent test & evaluation – Rapid/quick response capability – Mandated in-house performance responsibilities – Provide large/unique R&D facilities not commercially feasible	– Yardstick or smart buyer – Advanced capability in areas of limited interest to private sector – Rapid/quick response capability – Provide large/unique facilities not commercially feasible – Infuse "art of the possible" into defense planning – Provide full-spectrum capability	– Yardstick or smart buyer – Infuse "art of possible" into defense planning – Act as principal agent in maintaining tech base – Avoid technological surprise & ensure technological innovation – Support the acquisition process – Provide large/unique R&D facilities not commercially feasible – Rapid/quick response capability – Be a constructive advisor for DOD directions and programs based on technical expertise – Support the user in the application of emerging and new technology – Translate user needs into technology Requirements for industry – Serve as S&T training ground for civilian and military acquisition personnel	– Lowest cost to the Sponsor – Improve planning and avoid technological Surprise – Rapid/quick Response Capability – Flexibility and Responsiveness – Inherently governmental tasks – Corporate memory – Technology and systems integration – Reducing management complexity – Continuity of Personnel and Facilities Across a System's Lifecycle – Long-term/low pay-off essential military R&D

Conversely, loss of this internal technical competence means loss of control over outsourced work, which can have catastrophic consequences for any business, public or private. Colvard cites several examples of what can happen when such technical capability is lost or technical advice ignored.[55]

- *ValuJet lost technical control of its fleet and was grounded after one of its jets crashed in the Florida Everglades in 1996. The company had contracted out all maintenance and lost the ability to recognize its technical troubles. Further, there are reports that the government inspector who monitored ValuJet was not technically qualified.*

- *NASA decided to go through with the doomed Challenger launch in 1986, despite technical advice to delay it because of cold weather's effects on the space shuttle's O-rings. The decision was managerial, not technical. It was reported that the contractor's regional manager suggested to the engineer who provided the technical advice that the company not appear uncooperative, since the contract was coming up for rebid. Barbara Romzek and Melvin Dubnick, authors of <u>American Public Administration: Politics and the Management of Expectations</u> (MacMillan, 1991), say "there has been a shift in NASA from a system of professional accountability, which emphasizes deference to expertise within the agency, to a management system incorporating bureaucratic accountability."*

- *The Navy lost its surface-launched missile engineering capability, at least for the short term, in a defense industry shakeout that followed the Cold War. General Dynamics Corp. operated the Navy Industrial Reserve Ordnance Plant in Pomona, Calif., for years. The organization ultimately moved to Tucson, Ariz., after being shifted from General Dynamics to General Electric to Raytheon. Many people who had worked for years building Navy missiles did not relocate.*

This incident with General Dynamics demonstrates why the DOD must retain the technical capability to assess the products it gets. Even if the company's move made sense from a shareholder's perspective, as Colvard points out, it left the Navy to rebuild a technical capability that had already cost taxpayers billions of dollars. Further, when an airline loses its ability to oversee the technical product it is getting from its suppliers—perhaps with devastating consequences—flyers can at least turn to another airline. The DOD, however, may not be able to turn to other suppliers because of the recent, extensive consolidation of the defense industrial base, which has left the Department with few competitors in many important areas.

The second space shuttle disaster also likely occurred, in part, because of loss of technical control over an outsourced product. In this case, Boeing shifted space shuttle engineering offices from California to Texas. Unfortunately, some 80 percent of the 500 employees refused to move, forcing Boeing to hire new employees, including many engineers. The STS-107 shuttle mission was the first one for which the new Texas office had primary responsibility. Engineers who did not move to Texas argued after the disaster that they would have reached a different conclusion about the damage the foam impact might have caused. If so, the loss of corporate memory resulting from the transition could have partly caused this tragedy.[56]

Extensive outsourcing has in fact begun to affect the DOD's bottom line, causing a reaction that underscores the necessity of the laboratories' yardstick role. As one author puts it, *"the trials and tribulations experienced in various Pentagon big-ticket programs in recent years have prompted a thorough self-examination at the Defense Department. At issue is who really is to blame for failures, cost overruns and an overall dearth of innovation."*[57] The author further points out that the DOD is rethinking its management approach, *"so it can become a 'smart buyer,' better equipped to oversee increasingly complex technologies, and to determine if a potentially innovative technology is worth the financial risk."* Contractors' overemphasis on the bottom line may in fact help account for problems in such programs as space and information technology. But these problems *"can also be attributed to a gradual decline in the Defense Department's in-house expertise to manage and oversee highly intricate weapon systems and vast network integration efforts."*

In "The Case Against Privatizing National Security," Ann Markusen, a university professor and member of the Council on Foreign Relations, also discusses how and why many defense officials are beginning to reexamine their faith in outsourcing:

[Privatizing] military research and development is especially problematic. Traditionally, the United States has maintained strong in-house research and development laboratories to work on sensitive and pressing technical issues. These laboratories have contributed to the technological superiority of the American military…. They have acted as reservoirs of expertise than can be used to oversee, evaluate, and compete with private-sector research and development efforts, a "yardstick" function that economists and defense experts have always considered *appropriate for government. Yet in 2000, the Defense Science Board recommended that the services hire scientists and engineers 'from universities, industry, and nonprofits for a majority of the professional staffs of the defense laboratories.' This is considered dangerous by insider critics, because it will degrade the ability of the defense laboratories as performers of research and as evaluators of for-profit research performance [leaving] the military dependent upon advice that is not insulated from commercial interests.* [58]

A final indication the privatization trend may be winding down is that the revivalist fever it engenders is sure to abate as it always has in the past, every time the evangelists of outsourcing manage to reap a new wave of temporary converts. This is especially the case once the shortcomings and excesses of the new movement become apparent. In fact, an examination of the historical record discloses that today's faith in privatization is just another swing in a recurring historical cycle. [59] As has been noted, the pendulum may already be once again swinging back in favor of rebuilding DOD's capacity to do more in-house work.

2.7 Reluctance of Defense Companies to Invest in Long-Term/High-Risk Research

A second major factor that argues in favor of increased in-house performance of S&T in certain areas of defense is a growing unwillingness in the private sector to engage in technical work that involves long-term and/or high-risk investments of their own money. Commercial companies, including many in the technology sector, are primarily interested in a quick return on investment to boost profits and please shareholders. This is equally true as regards the defense industry, where large companies are focusing their dwindling in-house R&D efforts on things like risk reduction and cost containment, with little of their own money going toward developing innovative technologies. [60]

A report by Booz Allen Hamilton discusses in some detail this trend of disinvestment in research by defense companies. [61] It points out there are basically three sources of R&D funds for the defense industry:

- The U.S. Government for funded development programs (contract research and development)
- Company-sponsored research and development
- Independent Research and Development (IR&D) paid for by the government, but spent at the discretion of the contractor

With respect to the third funding source—IR&D (also abbreviated as IRAD)—the report points out a few problems, including a growing national security risk. First, "*discretionary IR&D funds are becoming less discretionary; simply put, the 'I' in 'IR&D' is slipping away.*" Second, IR&D is increasingly "*aligned toward near-term programs or used to warrant the development of a specific deliverable rather than long-term independent research and development. While the near-term risks to the contractor haven't increased, the long-term risks to the U.S. and its citizens have increased even more.*" In 1996 for example, 75 percent of the IR&D investments by space companies fell in the discretionary category, a figure that declined to just 23 percent by 2000. Instead, more IR&D was directed toward near-term programs—from 20 percent in FY 1996 to 45 percent in 2000—and toward developing a specific deliverable—from just 5 percent in FY 1996 to 32 percent in FY 2000. Long-term independent R&D decreased proportionately.

Not only has defense industry R&D investment become less discretionary and more devoid of risk, it has also dwindled in overall size. For example, over this same period (1996 through 2000), company-sponsored R&D investment (the first two categories mentioned above), as a percent of several large defense and aerospace companies' sales, fell from 4.1 to 2.9, a decline of almost 30 percent.[62]

Some argue that the fall-off in company IR&D resulted in part from changes to the 1971 law that allowed companies to recover their IR&D expenses as general and administrative overhead to build their technology base. A recent report by the Potomac Institute for Policy Studies argues this point:[63]

Companies with IR&D programs in excess of $4 million were required: (1) to submit a technical plan describing each technical project, which the DOD would evaluate for "potential military relevance"; (2) to negotiate an agreement with the DOD which established an IR&D ceiling for recovery...; (3) to present an on-site review of its IR&D program to the DOD at least once every three years. In December 1991, responding to pressure from industry to simplify the process and to allow full recovery of IR&D expenses, Congress passed PL-102-190, which stated that "independent research and development and bid and proposal costs of DOD contracts shall be allowable as indirect costs on covered contracts to the extent that such costs are allocable, reasonable, and not otherwise unallowable by law or under the Federal Acquisition [Regulations] (FAR)."

This change allowed companies to fully recover their IR&D expenses, but it also negated the requirement for IR&D ceiling negotiations with DOD, and for industry reporting and DOD review and oversight. According to the Potomac Institute report, two important but unintended consequences resulted:

First, it allowed (thus effectively encouraged) industry to reduce overhead rates by reducing IR&D investments. IR&D budgets shrank to less than half their levels before the law change, and the character of IR&D work became nearer-term (more like bid and proposal funding).... Second, the DOD and industry lost overnight a forcing function to encourage and assure access by individual performers to each other's research and development, ending a period of many years of mutually advantageous technical communication and leverage between the public and private sectors.

Regardless of the cause, the defense industry is investing less of its own money in new, innovative technologies. Unless the trend is reversed, defense companies will be less and less able to provide solutions the DOD requires, with the result that the Department's in-house laboratories and centers may become an increasingly attractive alternative.

2.8 Impact of Defense Transformation on Defense Industrial Base

A third factor that may well impact in-house versus out-of-house performance of defense basic and applied research is DOD's on-going defense transformation effort. At issue is the following question: how will transformation affect the ability of the major defense companies to provide new technologies?

Some observers are beginning to suggest the shift away from the acquiring major weapon systems and toward developing new technologies that address terrorist threats may imperil these companies' long-term survival. For example, some defense officials and analysts believe that proposed funding cuts in such major programs as Lockheed Martin's F/A-22 stealth fighter and Northrup Grumman's shipbuilding and repair programs may foreshadow other cuts to big-ticket programs.

One defense industry source quoted in a recent story on this subject said such changes "raise the question of how large defense contractors will stay in the game as the Pentagon puts less emphasis on buying big platforms such as aircraft carriers and stealth fighters in quantity and more on technologies designed to meet emerging security threats, including countering bioterrorism, developing new non-lethal and kinetic energy weapons, and fostering joint service science and technology efforts."[64]

The potential impact of defense transformation is also the subject of another recent article that examines how changing defense investments may fundamentally reshape the defense industrial base.[65] One source quoted is Suzanne Patrick, Deputy Undersecretary of Defense for Industrial Policy, who believes *"we'll have a completely different set of actors…in terms of corporations that we will draw on…. Of the current companies that exist, there may be a modest subset of the primes that still will be recognizable."* Perhaps more ominously, she also predicts that two or three of these companies *"will go belly up,"* while three to five *"may change quite dramatically, getting into other activities and tasks"* that suit the soldiers' needs.

It remains to be seen whether the small number of major defense and aerospace companies we have today will weather the change promised both by budget cuts and budget restructuring in response to defense transformation. If, however, they falter, where will the DOD look for the technology products it relied on them to produce? Again, one source may well be its own in-house laboratories and centers.

2.9   Limitations on Academia's Performance of Defense Science and Technology

As discussed early in this chapter, historically the DOD has relied on academia to perform much of its basic and applied research. In fact, academia performs the largest share of the Department's overall defense basic research program. According to National Science Foundation (NSF) data, U.S. universities and colleges performed 56 percent of all federally funded basic research in 2002.[66]

There are, however, reasons to believe it may become more difficult to get universities to undertake defense work, at least in certain areas. If so, alternative sources will be required, and these could include DOD's in-house laboratories and centers. There are several reasons for this postulated change, but most can be attributed to the security-conscious environment growing out of the events of September 11, 2001.

Even before "9/11," there were several long-standing constraints to unfettered performance of defense S&T by academia. Many of these were, and remain, self-imposed. For one, most colleges and universities prefer to focus on fundamental (basic and applied) research, the results of which can be published without restriction. For another, academia prefers to eschew classified work, which when done is often confined to off-campus facilities. The MITRE Corporation is an example of an off-site organization set up to do classified work related to defense research performed by the Massachusetts Institute of Technology (MIT) on behalf of the Air Force. Another historical constraint involves the cost of facilities and/or equipment which are, in many cases, so large as to prohibit private ownership.

After 9/11 the constraints on academic performance of research became even more restrictive, and significantly so. These new circumstances promise to curb the growth of academic performance of defense S&T in some areas by threatening the four principal values under with most universities operate. According to Eugene Skolnikoff, emeritus professor of political science at MIT, these values are "commitment to openness, resistance to classified research, maintaining open relationships between universities and industry (including foreign industry) and, of course, relations with foreign students." [67] Academia will likely react to threats to its long-held values by reducing research in certain areas, for example biological agents. Part of the difficulty stems from the fact that all technologies are dual-use. That is, any technology, regardless of the use for which it was created, can be turned to malevolent ends. Complicating the situation is the fact that many technologies are becoming both inexpensive and commercially available. This makes it easier for both state and non-state actors to acquire them for terrorist purposes. History demonstrates that almost any kind of technological knowledge will eventually leak out and proliferate throughout the world. And today, the rate of diffusion is greatly accelerated in a highly interconnected global community. Further, it is practically impossible to predict how a new technology will eventually be applied, especially in conjunction with other technologies that may enable a completely unforeseen application.

Skolnikoff further notes that in the post-9/11 environment, two diametrically opposed forces will likely complicate academic performance of defense research. One, most research universities are prospering, and in doing so are increasingly engaging in international activity. They educate more foreign students and carry out more collaborative efforts that involve foreign researchers, institutions, and companies. In fact, some academic departments would find it difficult or even impossible to operate without foreign students. At the same time, however, the U.S. is becoming more concerned over threats from abroad, and is seeking to limit some of the very activity universities are engaging in now more than ever.

The Congressional Research Service (CRS) has examined the possible impacts of counter terrorism actions on R&D and higher education. [68] They note that while there is widespread agreement on the need for such new measures, some could have unintended consequences. For example, many new restrictions aim to limit access to scientific and technical information. These include "*controlling access to research and development laboratories, self-policing, classification and reclassification of already released materials, withdrawal of information from Federal agency websites, possible additional exemptions from the Freedom of Information Act (FOIA), and withholding information categorized as 'sensitive but unclassified.*" There is also a growing list of new or proposed restrictions dealing with access to biological agents. These include "*proposals to register users of potentially toxic biological and chemical agents; to inventory laboratories that conduct research using pathogenic biological agents; to limit access to research and development laboratories and biological research agents; and to give tax preferences to firms that deal with bioterrorism.*" Some specific unintended consequences are the following: "*high financial costs, especially to academic laboratories, of instituting security and tracking measures, the possible deleterious impacts on freedom of scientific information exchange and scientific inquiry, and the possible loss to the United States of foreign technical workers in areas of short supply among U.S. citizens.*"

The issue of foreign graduate students and national security is another factor affecting academia's ability to do defense-related research. This is all the more true because, while American students are rejecting graduate study in mathematics, engineering, and the physical sciences, the numbers of international graduate students in these areas has increased. Today, for example, in the U.S. more engineering doctorates are awarded to international than domestic students. According to NSF data, the number of new U.S. doctorates earned by students on temporary visas rose from about 4,300 in 1986 to about 8,000 in 1991, a figure around which it has fluctuated for a decade. Significantly, foreign students, both temporary and permanent visa holders, earn a larger proportion of doctorates than at any other degree level.[69]

Aside from the question of whether foreign students remain in the U.S. and contribute to our S&T resources, there are concerns about admitting large numbers of them to American universities and allowing many of them to stay after graduation. Issues surrounding this controversy burst into the open in the spring of 2002, when DOD proposed a new policy aimed at the handling of unclassified research in both DOD and private-sector laboratories.[70] According to *Science*, the proposed policy "*would have Pentagon program managers decide if DOD-funded studies at universities, companies, or military laboratories involve critical research technologies, or critical program information. If so, the institutions and researchers conducting the work would have to prepare detailed security plans, label documents as protected, obtain prior review of publication and travel plans, and decide whether to place restrictions on any foreign scientists involved in the project."[71] Many of the areas that such a policy would affect are precisely those in which additional talent is needed.*

As with the effects of defense transformation on major defense companies, the effects of these changes remain to be seen. They are, however, further reasons the DOD should retain a highly trained S&T workforce capable of performing work in areas academic researchers no longer pursue.

2.10 Changes in Center S&T Capability Required by Globalization

Globalization[72] is a much talked about subject, especially its impact on the ability of the U.S. to remain a world leader in innovation. Discussions of globalization and technological innovation often include three aspects: *technological generation*, *technological exploitation*, and *technological collaboration*. All of these are important in the context of this book, because globalization of sources of new research and knowledge will place new demands on the S&Es in the DON warfare/systems centers concerning their ability to track, assess, and, when appropriate, apply this rapidly growing knowledge to military problems. The challenge of globalization to American technical leadership is of growing concern to U.S. policy makers, especially its economic and national security implications.[73]

An important point to note regarding collaboration is that today, only a handful of firms and other organizations can innovate alone. More and more frequently, innovation requires a network of organizations working together. This is especially true in the case of the most valuable, knowledge-intensive, and complex technologies, such as computers, semiconductors, telecommunications equipment, aircraft, and biotechnology. Moreover, the ever more rapid dispersion of scientific knowledge around the world means an increasing percentage of what many refer to as *innovation networks* involves a mix of global partners. This subject is further discussed in Chapter 4, which addresses the need to fashion a new DOD S&T enterprise.

The rapidly accelerating accumulation of intellectual capital—including an advanced S&E workforce—in other countries has worrisome implications. Especially noteworthy is Asia's increasing homegrown technical capability, exemplified by the rapid growth in the number of students receiving S&E doctorates from Asian institutions.[74] Diana Hicks, chair of the Department of Public Policy at the Georgia

Institute of Technology, has compiled extensive data on the strengthening of research capability in Asian countries. She points out that the global landscape is changing because, during the past decade, *"many governments, convinced that their economic futures lay with knowledge-based economies, sought to strengthen national research and education."* Further, *"Increased foreign scientific competitiveness may be little noticed from within the U.S...whose output still dwarfs that of any other country. Nevertheless, in aggregate these shifts are beginning to have an impact on U.S. research."*[75] In other words, these countries are acquiring high-end innovation capabilities by purposefully focusing their investments in R&D and technical talent. Significantly, it is not just a matter of other countries building their capabilities; it is the development of very broad technical infrastructures that matters in the long term.

Already, these trends are affecting U.S. research outputs relative to the rest of the world. The "Task Force on the Future of American Innovation," which examined this issue in detail, points out that the U.S. share of S&E papers published worldwide declined from 38 percent in 1988 to 31 percent in 2001, with Europe and Asia being responsible for the bulk of recent growth in scientific papers.[76] In fact, the Task Force notes that Western Europe's output passed the U.S. in the mid-nineties, and Asia's share is rapidly growing. Moreover, from 1988 to 2001, *"the U.S. increased its number of published S&E articles by only 13 percent. In contrast, Western Europe increased its S&E article output by 59 percent, Japan increased by 67 percent and countries of East Asia, including China, Singapore, Taiwan, and South Korea, increased by 492 percent."*[77]

The downward trend in U.S.-authored scientific and technical articles is evident in most fields, with the greatest decrease occurring in engineering and technology articles (down 26 percent between 1992 and 1999).[78] Other declines over this period included articles in mathematics, physics, chemistry, and oceanography, all of which are important to the DOD. Figure 2.13, which is taken from a slide in an April 2005 briefing by the DDR&E illustrates a striking example of the increase in publications in physics from countries other than the U.S, a field of critical import to the DOD.

Perhaps one of the most important points about the globalization of technology is the growing offshore accumulation—"off-shoring"—of intellectual capital and industrial capability in many important technologies. If not dealt with soon, such trends may well threaten the ability of the DOD to maintain its technological lead over adversaries.

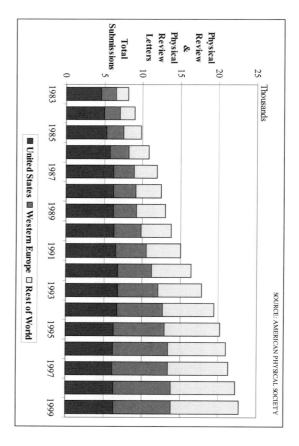

Figure 2.13: Trends in the Publications of U.S. Physics Papers (Source: April 2005 Briefing by DDR&E Base on American Institute of Physics Data)

Consider just one example—technological globalization's threat to the electronics sector. According to an unpublished report by the distinguished Pentagon Advisory Group on Electron Devices (AGED), the "Department of Defense faces shrinking advantages across all technology areas due to the rapid decline of the U.S. electronics sector....off-shore movement of intellectual capital and industrial capability, particularly in microelectronics, has impacted the ability of the U.S. to research and produce the best technologies and products for the nation and the warfighter.... DOD is forced to rely on perceived system integration advantages to maintain superiority."[79] The group argues this could force the DOD to obtain the most advanced technologies from overseas, a situation that could assign those nations both political and military leverage. According to the AGED, "In the area of battlefield communications and data networks, the global availability of wireless communications and high data rate fiber optical landlines has greatly reduced this advantage even against the less sophisticated terrorist threat. Use of best commercial chips and processors levels the playing field for allies and adversaries."

Off-shoring is accelerating daily. The manufacturing sector was the first to experience it, beginning in the 1980s and 1990s. However, the trend began to impact the service sector soon after with customer service centers and other service-oriented functions being handed off to India and other Asian countries. Engineering and medical services are just two areas where this trend has had a major impact.

More recently, however, a growing number of U.S. companies are farming out their R&D—indeed, their very ability to innovate—to off-shore entities, ostensibly to cut costs and get their high technology products to market faster. For example, major firms such as Dell, Motorola, and Philips are now buying complete designs for many of their digital devices from Asian developers, then *"tweaking them to their own specifications, and slapping on their own brand names."* [80]

A related consequence is that while many U.S. companies are downsizing at home, they are boosting hiring at their laboratories in India, China, and even Eastern Europe. This drains high-tech investment capital away from the U.S. and into these countries, a point Battelle makes in its most recent R&D funding forecast: *"the U.S. industrial community will be strained to invest in U.S. R&D as China, India, and other Asian economies develop their own technological capabilities and draw off investments to support their own burgeoning markets that might normally go to U.S. facilities."* [81]

Unquestionably, the line that divides commodity work and R&D is sliding year by year. "The implications for the global economy are immense. Countries such as India and China, where wages remain low and new engineering graduates are abundant, likely will continue to be the biggest gainers in tech employment and become increasingly important suppliers of intellectual property." [82]

As a result, there is a growing consensus among U.S. government and many business leaders that off-shoring threatens economic and national security. A recent report by the PCAST voiced concerns that *"while not in imminent jeopardy, a continuation of current trends could result in a breakdown in the web of innovation ecosystems that drive the successful U.S. innovation system."* [83] That is to say, while a snapshot might suggest all is well, the trends tell a different story—there is no room for complacency. The story told by the "slope" of the trend lines suggests a currently stable situation may well be heading into one of instability. This is a point developed in detail in Chapter 3.

2.11 A Call for a New DOD S&T Enterprise

Changes in the environment in which its laboratories and centers will be operating in the 21st century have begun to undermine the approaches the DOD previously used to maintain technological superiority over our adversaries. The rapid spread of capabilities derived from new technologies, now widely available on a global basis, raises important new questions. For one, how can the DOD mine and employ these technologies to maintain its technological dominance? For another, will the DOD's current approaches to the process of technological innovation, especially those that appertain to the D&I phases, sustain us in the 21st century?

Clearly, more in-house S&T capability than we have today will be needed. In fact, it will have to be increasingly sophisticated for the DOD to remain a peer-player on the global technology scene. Regrettably, however, just as these demands are growing, the very S&T workforce the DOD needs is dwindling and in urgent need of renewal, especially in light of its recent deterioration. Indeed, some policy-makers have recognized this need, and proposed new initiatives aimed at bolstering the S&T workforces in the in-house laboratories and centers, but these are far from sufficient.[84]

It is nearly axiomatic that an organization remains "world-class" by hiring and retaining productive, high-quality people, including a few—the top 10-percent—who have exceptional talent. This is especially true of cutting-edge S&T organizations. Thus, if the DOD S&T enterprise is to remain world-class, it needs the flexibility to do whatever is necessary to hire, train, and retain a cadre of the best and brightest scientific and engineering talent available—world-class talent.

What is needed is a fresh look at the entire innovation process, and in particular, the role of the DOD's in-house laboratories and centers and their workforces. What is really called for is a new DOD S&T enterprise, a subject discussed later.

Chapter 3

From Global to Local: Looking Behind the Numbers

3.1   Introduction

As stated earlier, a principal purpose of this book is to highlight the continuing importance of S&T to DOD in ensuring U.S. military forces retain their current technological superiority over any and all adversaries. Regrettably, the ability of DOD's laboratories and centers to perform their S&T roles is at risk today largely because of what has happened to the workforce they employ to carry out that work. The previous chapter mentioned several factors that threaten the continued vitality of this workforce such as its dwindling size and rapid aging.

Some of the trends that affect the defense S&T enterprise and its in-house technical workforce are local to the DOD. Examples include S&T funding trends, outsourcing of technical work, and the impact of past BRAC actions on the human capital in the laboratories and centers. These and other topics are discussed elsewhere in this book. There are also trends that are national in scope, such as the reluctance of young Americans, especially women and minorities, to pursue science and engineering careers. There is a large body of literature that has examined U.S. education issues as well as those that relate to the quality of primary and secondary (K-12) education in the U.S., and readers who are interested in such issues should consult those sources. Much has also been written about U.S. graduate and post-graduate education challenges. In this regard, a later chapter will address several DOD/DON educational initiatives aimed at increasing interest among U.S. students in science, technology, engineering, and mathematics (STEM).

Besides local and national trends, there are others that are global in scope. Some of these have been mentioned previously, and they primarily relate to the rapid accumulation of intellectual capital and technical infrastructure in other parts of the world. Especially noteworthy in this regard are trends in several Asian countries, with China being of particular concern.

Global trends too have been widely discussed and debated, especially in the context of their impact on U.S. economic world leadership.[1,2,3,4,5] However, less attention has been paid to what these trends have to say about U.S. national security vis-à-vis other countries, or how they might impact this country's future defense posture. The fact is that some of these trends will have a significant influence on the kind of S&T enterprise the DOD will need to meet its 21st century warfighting needs, and this will be discussed in Chapter 4.

This chapter will examine a few of the global trends that directly impact the size of the talent pool from which the DOD must recruit its future S&T workforce, because it is important to understand the factors that determine whether this pool will be adequate for future U.S. national security needs. This chapter will look at trends related to U.S. technical output, such as the number of scientific articles published in peer-reviewed technical journals. This subject was briefly mentioned in the previous chapter, where it was noted that much of the decline in numbers of publications has occurred in S&E fields of importance to the DOD, e.g., mathematics, physics, chemistry, and oceanography. In this chapter, the intent is to "look behind the numbers" in two representative areas of significant importance to future defense efforts to see what DOD and national security implications they might have. These areas are nanotechnology and EMs. The discussion here will also describe a causal model that addresses a number of trends that will be used to illustrate the complexity of predicting future S&E production trends and needs, and the fact that much of the data needed to populate such a model have not been collected, certainly not in a methodical and consistent way.

3.2 Global Trends and the U.S. S&E Labor Market

It has already been noted that the issue of whether the U.S. has enough, or the right kinds of S&Es to maintain its S&T technological leadership in the world is controversial, and that there are proponents on both sides of the debate. To the extent that human capital in the fields of S&E is important for economic growth, this issue also has to do with U.S. economic leadership in the world and how globalization could affect that leadership.

One thing is clear: a number of global trends have a direct bearing on the U.S. S&E labor market and, ultimately, the supply of S&Es available for employment by DOD's laboratories and centers. Richard Freeman, the Harvard University economist, identifies two of the most important.[6] First, he notes that by 2001-2002, the U.S. share of students worldwide enrolled in tertiary education was 14 percent, less than half the proportion in 1970, a figure that contrasts sharply with rising numbers in Europe and Asia, especially China. In addition, as late as 1975, the U.S. granted more S&E Ph.D.s than Europe, and more than three times as many as all of Asia. Now the European Union (EU) countries graduate about 50 percent more than the U.S. and Asia has edged slightly ahead of the U.S.

The second trend discussed by Freeman has to do with the decreasing attractiveness of S&E to U.S. students. In short, many U.S. S&Es are not all that well paid considering their highly-specialized skills. This is a fortiori the case as regards those with Ph.D. degrees because of the considerable length of time required to obtain that level of education. Too often, the result is that many of the brightest U.S. students turn instead to other more lucrative fields such as business, law, and medicine, which often require less training, and where the payoff is often much greater and comes sooner. These labor market trends are important because they directly affect the availability of S&Es with advanced degrees to work on S&T problems important to the DOD and the military services.

In the past, American dominance in most fields of S&E was so overwhelming that policy changes related to the U.S. technical infrastructure could be taken without any great regard for their global implications, particularly how they would affect technological competition from other countries. However, this situation has changed rapidly in the last decade as other countries have begun to build up their own technical infrastructures and pools of talented S&Es. As a result, many foreign students who once considered American universities as their only choice for studying S&E now find they have equally attractive options in other countries, including, in a growing number of cases, their own. To the extent there is a significant diminution in this pool of foreign talent, it will be reflected in the size of the pool of S&E talent available to the U.S. and, consequently, to the DOD.

Such workforce issues and their national security implications were the primary focus of a recent Engineering Dean's Council Public Policy Colloquium held in Washington, D.C. The colloquium, "Engineering's Role in the Nation's Future," involved a group of more than 100 deans of engineering schools who heard a number of speakers address this subject. Congressman Jim Cooper (D-TN), who gave the keynote address, expressed growing worry over the future of American competitiveness in the world. Among other points, Cooper noted that "In today's business world, a strong foundation in science and engineering is vital to success." In this regard, he noted that "Enrollment in science and engineering graduate programs among American students dropped 10 percent between 1994 and 2001, while it rose 35 percent among international students. We clearly have to do a better job of encouraging America's young men and women to seriously pursue studies in the fields if we want them to succeed and we want the American economy to continue to lead."

As Cooper and many other individuals and groups have noted, the U.S. is still in the lead in most areas that are important to American competitiveness. However, just as a snapshot taken half way through a horse race is not a good predictor of which horse will ultimately win the race, a snapshot of today's global trends will not reveal where the U.S. may be headed in the future. A better understanding of our future direction as a nation can only be glimpsed by collecting and analyzing relevant trend data. There are many sources of such data; however, the NSF has one of the most comprehensive and authoritative collections, and a look at these data reveal a number of important facts, some of which are briefly discussed in the following paragraphs. These data can be found in the NSF compilation referred to herein as Indicators 2004.[7]

While the number of college age students in the U.S. has been fairly steady over the last 25 years, and is projected to remain steady for the next several years, competing countries such as China and India have college age populations that are very much larger than that of the U.S. (several hundred percent larger). This population advantage alone provides a much larger pool of potential students to attend college and pursue S&E degrees in these countries. In fact, Freeman suggests that the greatest threats to U.S. economic leadership will come from China and India in large measure because they are the world's most populous countries. This population advantage means that even if only a small proportion of their workforces are dedicated to S&E, the absolute numbers of S&E workers will still be large and will feed their R&D capabilities. They will also enjoy a significant labor-cost advantage over the U.S. in the S&E field for at least the foreseeable future, a factor that makes them a highly desirable target location for technology firms involved in global trade. In addition, in several foreign countries the number of students in the 20-24 age cohort choosing to attend college exceeds that of the U.S. For example, the ratio of degrees granted to 24 year olds in several Organizations for Economic Cooperation and Development (OECD) countries (i.e., Australia, New Zealand, Netherlands, Norway, Finland, the United Kingdom, and France) exceeds that of the U.S. Furthermore, in a growing number of countries, a larger proportion of college students study S&E than in the U.S. This stems in part from the growing unattractiveness of S&E occupations for many American students, a point previously noted. Figure 3.1 compares first university S&E degrees in Asia and Europe to those granted in North America. It shows that both Asia and Europe produce more engineering degrees than does the U.S.

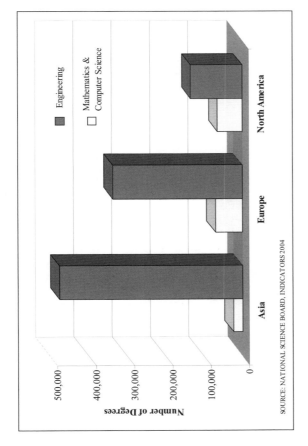

SOURCE: NATIONAL SCIENCE BOARD, INDICATORS 2004

Figure 3.1: First University Degrees for Different Regions by Field: 2000

According to the NSF, in the year 2000 some 17 percent of all university bachelor-level degrees in the U.S. were in the natural sciences and engineering, compared to a world average of 27 percent and to 52 percent in China. For the past three decades, S&E degrees have made up about one-third of U.S. bachelor's degrees. Corresponding figures were higher for China (59 percent in 2001), South Korea (46 percent in 2000), and Japan (66 percent in 2001). The contrast is even greater in engineering fields where, compared to Asia and Europe, the U.S. has a relatively low proportion of S&E bachelor's degrees in engineering. For example, in 2000, students in those two regions earned 40-41 percent of their first university S&E degrees in engineering compared to about 15 percent in the U.S.[8]

In the NSF data, S&E degrees include degrees in Social and Behavior Sciences. In 2000, 1,253,121 first university degrees were awarded in US. Out of these, 398,622 were S&E degrees. Out of the 398,622 S&E degrees, 188,188 were in the Social and Behavioral

Sciences. Hence, overall 32% (398,622/1,253,121) degrees were awarded in S&E fields in US. However, in order to correctly compute the influence of these numbers on the US technological leadership, we should exclude Social and Behavioral Science numbers from S&E numbers. So, if we exclude Social and Behavior Science degrees from S&E, then 17% ((398,622-188,188)/1,253,121) degrees were awarded in Natural Sciences and Engineering (these degrees include natural, agricultural, and computer sciences, mathematics, and engineering).

One conclusion that can be drawn from these data is that students in many Asian countries consider an education in S&E as a very desirable career path to success. This is certainly true in China, where engineering is highly valued as a career objective. There, majors in engineering account for some 3.7 million students. Indeed, former President Jiang Zemin, current President Hu Jintao, and every member of the nine-man Central Committee of the Communist Party of China are engineers by profession, as are scores of government ministers, governors, chief executive officers and entrepreneurs. This situation stands in stark contrast to the situation in the U.S., where many, if not most, government leaders are trained as lawyers and businessmen. In fact, at present, less than 1 percent of the members of the U.S. Congress have a science or engineering background.

Baccalaureate degree production is important because it feeds the graduate level pool of students, and ultimately impacts the number of Ph.D. graduates in S&E. These trends in first university degrees strongly suggest the number of Ph.D. degrees granted in S&E outside the U.S. will rise sharply, whereas the number granted in the U.S. will likely remain relatively stable for some years to come. In fact, countries such as China that granted few Ph.D. degrees in 1981 increased their number by over 61 times (from 125 degrees in 1985 to over 7600 in 2001). Taiwan and South Korea have also shown impressive increases in doctoral degrees, five-fold and nine-fold respectively, Figure 3.2.

As noted, these countries have a large student–aged population that is growing, and many will continue to pursue S&E fields. This trend will be further stimulated by the rapidly-growing movement on the part of U.S. corporations to off-shore technical work to these countries. This trend will help fund both the growth of the economies of these countries and that of their technical infrastructure. Because of its growing importance, the impact of off-shoring on the future S&T enterprise is discussed further in Chapter 4.

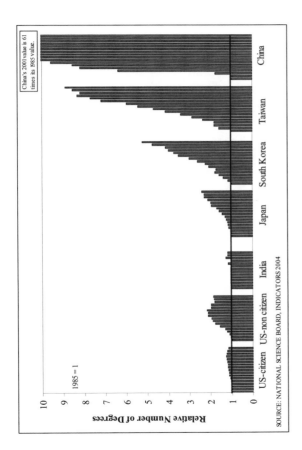

SOURCE: NATIONAL SCIENCE BOARD, INDICATORS 2004

Figure 3.2: Growth in S&E Doctoral Degrees for Various Countries: 1985-2001

Figure 3.3 compares the growth rate in S&E Ph.D. degrees awarded by field and citizenship of recipients. These data show declines in the number of doctoral degrees in science, engineering, and mathematics/computer science. The data also show increases in the number of foreign citizens earning S&E degrees. The strong growth in S&E doctoral degrees awarded by other countries is a further indication of the kind of changes that may be expected to occur in the future as other countries' economies and educational systems develop to a level that can more equally compete with the U.S. As a result, it can be expected that fewer and fewer foreign-born students will feel the need to come to the U.S. in order to obtain a Ph.D. degree in many areas of S&E. This trend will also affect stay rates and, ultimately, the pool of S&Es available for all kinds of work in the U.S., including employment in the defense sector.

A closer look at NSF data shows that students from just a few foreign countries accounted for nearly 70 percent of all foreign recipients of U.S. S&E doctorates from 1985 to 2000. Asian countries sending the majority of doctoral students to the U.S. have been China, Taiwan, India, and South Korea, in that order. Major European countries of origin have been Germany, Greece, the United Kingdom, Italy, and France.

Figure 3.3: Growth in U.S. Doctoral Degrees Earned by Field and Citizenship: Selected Years, 1977-2001 (Data has been normalized using 1977 as the base)

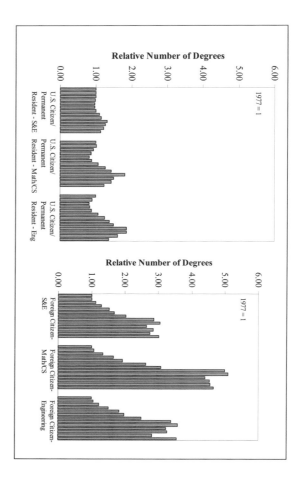

In the case of Taiwan, in 1985 their students earned more U.S. S&E doctoral degrees than students from India and China combined. The Taiwanese number of degrees increased rapidly over the period of 1985 to 1994 (from 746 in 1985 to 1,300 in 1994). However, as Taiwanese universities increased their own capacity for advanced S&E education in the 1990s, S&E doctorates earned from U.S. universities by Taiwanese students declined to 669 in 2000. Similar trends can be expected in the case of other Asian countries.

Another interesting trend has to do with the citizenship of the majority of S&E post-doctoral students (postdocs) at U.S. universities. Here, NSF data show that in the 1970s there was a large difference between the number of U.S. postdocs and foreign postdocs at U.S. universities: most were U.S. This gap slowly closed over the next 10 to 12 years and remained small until the late 1990s, when the number of foreign postdocs began to increase, while the number of U.S. postdocs began to decline. This situation is likely related to the decrease in U.S. S&E doctoral degrees and the corresponding increase in foreign S&E doctoral degrees shown in Figure 3.3.

3.3   U.S. Technical Output Trends

The growing scientific and technical infrastructures in Europe and Asia have already impacted U.S. leadership in terms of such metrics as the numbers of academic papers published in peer-reviewed technical journals, citations of those papers, number of patent applications, number of patents granted, etc. Citations—the number of times a paper or patent is referenced by other researchers or in other patents—are another indicator of technical output of a country. In particular, citations provide an indication of the paper or patent's relevance and importance to the scientific world. Since it often takes years for some patent applications to be approved and the actual patent granted, the number of patent applications by a country may be considered a more responsive or leading indicator of changes in U.S. technical output than patent awards themselves.

In terms of output, NSF Indicators 2004 data show that the number of U.S. scientific publications has remained essentially flat since 1992, while output has grown strongly in Western Europe and several East Asian countries. In fact, the number of scientific papers published by researchers in the Asia Pacific region could exceed the number from the U.S. within six or seven years. According to Science Watch,[9] the U.S. slide in output has been particularly noticeable in the areas of physics, engineering and in materials science. In fact, the *"Asia Pacific concentration in the physical sciences is even clearer upon examination of the detailed subfield covered in National Science Indicators. Between 2000 and 2004, authors from this region were most heavily represented in the subfield designated Materials Science & Engineering, having a hand in 54,754 papers, or 42.12% of the 130,004 Thompson[10]-indexed papers published in that field during the five-year period."*

Figure 3.4 depicts the number of S&E articles generated by various countries and regions of the world over the period 1988-2001. NSF data indicate that in 1998, the production of Western Europe reached the same level as that of the whole of North America and has kept pace since that time. The number of articles produced by Asian countries has also shown rapid growth, with an almost 40 percent increase in article output over the period 1988 to 2001.

Despite the global trend in publications, the S&E literature of the U.S. is most widely cited by non-U.S. scientists. However, the volume and world share of citations of U.S. S&E literature have been declining as citations of S&E literature from Western Europe and East Asia have increased. Trends in citation patterns by region, country, scientific field, and institutional sector are indicators of the perceived influence and productivity of scientific literature across institutional and national boundaries. On the basis of volume, the major producers of scientific articles—the U.S., Western Europe, Japan, and other OECD countries— are those whose S&E literature is most cited. According to NSF *Indicators 2004* data, "*In 2001, the United States' share of the world's output of cited S&E literature was 44 percent, the largest single share of any country. Collectively, the OECD countries accounted for 94 percent of the world's cited scientific literature in 2001 a share that exceeded these countries' share of the world output of S&E articles.*" In addition to having the most widely cited S&E literature, the U.S. also has the largest share of internationally-authored papers and it collaborates with the largest number of other countries. But, here again, its lead has declined. Other indicators of the technical output of a country include patent applications, patent awards, and citations in U.S patents to S&E literature. Consider the example of U.S. patents granted by residence of the inventor, Figure 3.5. It shows that South Korea and Singapore have demonstrated the most rapid growth over the period 1989-2001, with an almost 13-fold increase and a 21-fold increase respectively in patent applications in the United States. Other impressive increases are evidenced by China, Taiwan, and India. Although the U.S. and Japan still lead in patent applications, their rates of growth are less impressive.

Figure 3.4: S&E Articles by Region and Country/Economy: 1988-2001

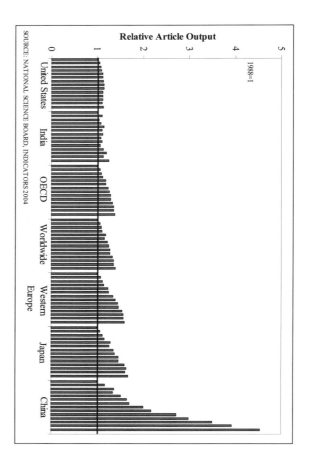

SOURCE: NATIONAL SCIENCE BOARD, INDICATORS 2004

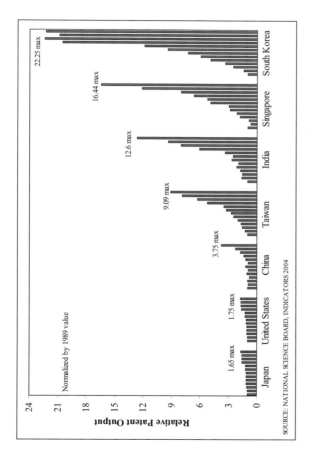

Figure 3.5: U.S. Patents Granted by Residence of Inventor: 1989-2001

SOURCE: NATIONAL SCIENCE BOARD, INDICATORS 2004

Another previously-mentioned indicator of changes in U.S. technical output is the number of U.S. patent citations to S&E articles. This is an interesting measure because it provides some indication of the linkage between research and its practical application. Although existing patents are the most often cited material, U.S. patents increasingly have cited S&E articles. In fact, according to Indicators 2004 data, the number of U.S. patent citations to S&E articles indexed in the Institute for Scientific Information's Science Citation Index[11] (SCI) rose more than 10-fold between 1987 and 2002. The SCI provides access to current and retrospective bibliographic information, author abstracts, and cited references found in 3,700 of the world's leading scholarly science and technical journals covering more than 100 disciplines. Note particularly that most of this growth of article citations in patents was centered in huge increases in the life science fields of biomedical research and clinical medicine.

Figure 3.6 shows how the U.S. has fared over the period 1995-2002 as compared to other countries and regions in terms of patent citations to S&E articles. The rapid increase in citations of S&E research by U.S. patents attests to the growing importance of science in practical applications of technology. Much of this growth has been driven by increased patenting of research-driven products and processes in the life sciences. The U.S. patents most commonly cite articles authored within the academic sector, again primarily the life science fields of clinical medicine and biomedical research. Industry was the next most widely cited sector. The life sciences, particularly biomedical research and clinical medicine, dominated nearly every sector, with from 67 to more than 90 percent of all citations, including those sectors that had prominent citation shares in the physical sciences earlier in the decade (industry and FFRDCs). They experienced significant declines in citations of articles by these fields, whereas their share of life sciences citations grew significantly.

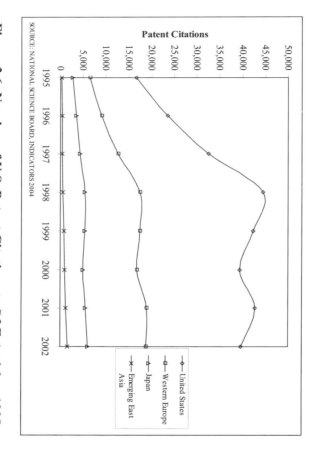

Figure 3.6: Number of U.S. Patent Citations to S&E Articles: 1995-2002

SOURCE: NATIONAL SCIENCE BOARD, INDICATORS 2004

The small sample of indicators presented above all support the view that the U.S. is still highly productive in terms of total scientific output. But they also suggest that the current lead the U.S. enjoys in terms of world technical and economic leadership may be eroding. There are, however, some voices that say this isn't necessarily a bad thing, at least in some areas. For example, Robert Samuelson, who writes frequently on economic issues, argues that another country's gain isn't necessarily our loss. To illustrate his contention, he points out, *"if a Swedish or Japanese company cured cancer or invented a super-efficient car,*

Americans would benefit quickly—just as Swedes and Japanese have benefited from technologies first developed in the United States.[12] Samuelson goes on to note that not every new Chinese or Indian scientist or engineer poses a threat to the U.S. because, as their economies grow, so does their need for technical talent to *"design bridges and buildings, to maintain communications systems, and to test products."* On the other hand, Samuelson notes, *"The dangers arise when other countries use new technologies to erode America's advantage in weaponry; that obviously is an issue with China. We are also threatened if other countries skew their economic policies to attract an unnatural share of strategic industries—electronics, biotechnology and aerospace, among others. That is an issue with China, some other Asian countries and Europe (Airbus)."*

Despite the words of solace offered by Samuelson and others, the rapid rise in the technological might of China is generating considerable concern in many quarters. Bruce Stokes has examined many of the pros and cons of the debate and provides a good assessment of both in a recent article in the *National Journal*,[13] where he had this to say: *"Few observers here believe that China is an imminent threat to U.S. global leadership in science and technology. But few doubt the Chinese government's high-tech aspirations or the trajectory of Chinese achievements in recent years. 'You never want to take a snapshot here,'* warned Michael T. Byrnes, president and chief representative in China for Tyco international. *'This is a moving-picture country.'"*

Many of those who dismiss concerns over the growing R&D infrastructures and workforces in countries such as China point to the fact that the U.S. not only leads in overall technical output, but also leads in terms of its quality. They also maintain that while globalization of the S&E workforce may weaken or, in the long-term, even undermine American economic leadership, the U.S. as a whole has a strong overall innovation system that will help slow this erosion. According to Stokes, *"The current consensus in Washington is that the best response is to run faster than the Chinese: Train more American scientists and engineers, spend more on R&D—and innovate, innovate, innovate."*

There is no doubt that adding up the numbers of papers and patents produced by a country will provide some overall picture of its current scientific and technological output. And, as has been seen, such a snapshot provides comfort to many here in the U.S. But in truth, such a picture does not offer much in the way of insight into how current trends may be impacting our future national defense posture. Put simply, if one only looks at total numbers of papers in technical journals, the U.S. is still in the lead at the moment. However, a closer examination of the U.S. portfolio of technical publications provides more interesting insights. For example, it shows that the portfolio is dominated by publications in medically-related life sciences (55 percent), while only about a quarter (24 percent) are dedicated to the physical sciences, and only eight percent to engineering, technology, and mathematics.[14] To a significant extent, this situation is the result of shifting priorities for Federally-funded research. Because of its importance, this is a topic considered in greater detail in the following paragraphs.

3.4   Impact of Shifting U.S. Research Funding

The growing research focus in the U.S. on the medically-related life sciences is largely the result of changing funding priorities of Federal departments and agencies. These changing priorities have produced significant shifts in the balance of funding among the various fields of S&E. A 1999 study[15] of research trends commissioned by the National Academies' Board on Science, Technology, and Economic Policy (STEP) found that several agencies spent less on research in 1997 than they had in 1993. (DOD was down 28 percent). Importantly, these agency reductions disproportionately affected most fields in the physical sciences (physics, chemistry, and geology), engineering (chemical, civil, electrical, and mechanical), environmental (geology, geophysics, oceanography, atmospheric sciences, ecological sciences), and mathematics, because these fields received most of their support from the agencies with reduced funding. The study found that Federal funding decreased by 20 percent or more between 1993 and 1997 in four fields: mechanical engineering, electrical engineering, physics, and geological sciences. The study, however, found growth in several areas, including computer sciences and medical sciences. In particular, it noted that most recent increases came in research fields supported by the National Institutes of Health, largely because Congress doubled that agency's budget from FY 1999 to FY 2003.

In 2001, the STEP Board published a follow-up[16] to its 1999 study of Federal research funding trends. It found that funding for the life sciences had increased to 46 percent of Federal funding for research in 1999, compared to 40 percent in 1993, while funding for the physical sciences and engineering decreased from 37 percent of the research portfolio in 1993 to 31 percent in 1999. Specifically, it found Federal funding in 1999 was still below 1993 levels for seven fields of research. Five of these fields--physics, geological sciences, and chemical, electrical, and mechanical engineering--were down 20 percent or more from 1993. The STEP Board concluded that a substantial shift had occurred in Federal funding, with significant declines in the physical sciences and certain fields of engineering, and substantial increases in the medically-related life sciences.

More recently, this issue has been the focus of the previously-mentioned PCAST, a group established to enable the President to receive advice from the business and academic communities on technology, scientific research priorities, mathematics, and science education. Its members are drawn from industry, education, research institutions, and other non-governmental organizations. The PCAST formed a panel on "Federal Investment in Science and Technology and Its National Benefits" to examine trends in Federal funding for R&D to determine their consistency with the nation's present and future needs. As part of its effort, the panel commissioned a study by the RAND Corporation to examine Federal support for R&D over the past 25 years and compare U.S. Federal and private sector R&D investments to those of our global competitors. The RAND study[17] and other information gathered by the panel were then used to develop the final report.[18] Its findings and recommendations include the following: Federal R&D funding relative to GDP continues to decline; private sector R&D investments are generally of a different nature than Federal support; and Federal funding for the physical sciences and engineering benefits[19] all scientific disciplines.

With regard to our national investment in R&D, the panel notes that 20 years ago, Federal funding for R&D exceeded that of private industry, but today the reverse is true. The panel notes that this is significant, because activities emanating from R&D investments that produced new growth have never been higher, including increasing numbers of patents and discovery disclosures. Indeed, there is strong linkage between federally-funded science and innovation. For example, a 1998 CHI Research study of the linkage of patent citations to the scientific literature found that, of patents granted to U.S. industry, approximately 73 percent of the science articles cited in the patent resulted from publicly-funded science.[20] The PCAST panel was not comforted by signs of increased private sector funding of R&D, noting: *"While strong support of R&D by private industry is to be commended, this source of funding cycles with business patterns and focuses on short term results by emphasizing development of existing technology rather than establishing new frontiers. Growing private investments in research do not replace the need for Federal support in certain critical areas and for long-term basic research, where the benefits cannot be measured in short cycles."*

PCAST summarizes the funding picture as follows: *"As a base point: in FY 1970, support for the three major areas of research, namely physical and environmental sciences, medically related life sciences and engineering was equally balanced. Today, the medically related life sciences receive 48 percent of Federal R&D funding compared to the physical sciences' 11 percent and engineering's 15 percent. Even if physical sciences, environmental sciences, math and computer sciences are combined, their total share is 18 percent."[21]*

According to the PCAST, the lack of funding in these disciplines, other than those that are medically related, is a cause of concern for a number of reasons. First, this has given rise to a situation in which both full-time masters and doctoral students in most areas of the physical sciences, mathematics, environmental, non-medically-related sciences, and engineering are decreasing. Over this same period, the numbers in the medically-related life sciences increased. Second, facilities and infrastructure in general for S&E are becoming less than adequate for meeting the challenges of today's research problems. Third, it is widely understood and acknowledged that the interdependencies of the various disciplines require that all advance together.[22] In other words, progress in such areas as the medically-related life sciences depends on continued progress in more fundamental areas, such as physics, chemistry, mathematics, and engineering. As an illustration of this latter point, PCAST points out that, at IBM, over 95 percent of the Ph.D.s who compose its nanotechnology research staff have degrees in the physical sciences and electrical engineering, areas in which graduate training is largely dependent upon support by the Federal government.[23] The increasing vitality and exciting discovery-initiating areas of interdisciplinary and multidisciplinary research and education cannot be sustained without investment in the non-medically-related basic sciences. Nanotechnology is only one example. The interdisciplinary research involving biological, information, nano- and cognitive (neuro-) sciences is moving rapidly. To sustain the extraordinary advances being made in these interdisciplinary areas, new collaborations in the fundamental, non-medically-related S&E disciplines must be nurtured now, not in some distant future. Indeed, the increasing complexity of advanced technology, which integrates multiple disciplines and technologies, depends on concurrent advances across many fields. The imbalance in America's scientific portfolio runs a serious risk of adversely affecting the capacity for innovation in a range of key sectors and impeding the ability to fulfill other critical national missions.[24]

This funding imbalance, with its heavy emphasis on the medically-related life sciences, has potentially serious implications for national defense. This point has been considered by the Council on Competitiveness, which suggests three national challenges that face the country: improved health care; energy and environmental quality; and national defense.[25] In each of these areas of national challenge, the

Council claims there are both a number of "contributing sciences," as well as a number of "enabling technologies," providing another perspective on the interdisciplinary nature of S&T as it is being done in today's world. Therefore, if we are to make headway in meeting national challenges in defense, the economy, and social stability, we must also do so in the contributing sciences and enabling technologies that underpin them, most of which are included among the fields that are, today, funded at considerably lower levels than in the past. So, again, we see that the physical, environmental, and non-medically-related sciences, engineering, and mathematics are pillars on which progress toward meeting national challenges stand.

The national challenge of defense (including homeland defense) is a major concern. The Council cites a number of contributing sciences and enabling technologies on which national defense depends. Contributing sciences include computer sciences, electromagnetic theory, materials sciences, physics, quantum mechanics, robotics, and transport physics. Enabling technologies include electronics, computing, the social sciences, human-interface technology, manufacturing technology, materials technology, nuclear technology, optical technology, and plasma technology.[26] Because of shifting national funding patterns, many of these areas are being reduced in absolute funding levels, with a potential negative impact on our ability to meet future national security needs. This funding imbalance will also have a negative impact on the medically-related sciences because many of the medical devices and procedures we benefit from today, e.g., endoscopic surgery, smart pacemakers, dialysis machines, imaging technologies (e.g., magnetic resonance imaging (MRI), Computerized Axial Tomography (CAT) scans and Positron Emission Tomography (PET) scans are the result of R&D in the physical sciences and engineering.[27]

3.5 Global Trends and Implications for DOD: Nano-S&T and Energetics Examples

As previously discussed, global trends are having a significant impact on the position of the U.S. as a leader in S&T output and to the extent these trends are not reversed, they will impact the ability of the U.S. to compete in a technological and economic sense on a world-wide basis. Also, as we have seen, a look behind some of the numbers reveals several significant points. First, our current leadership position in terms of published papers, patents, etc. is largely due to extensive R&D investments by both industry and the Federal government in the medically-related life sciences. Second, many areas of importance to the DOD are currently under-funded, and this has had an effect on both graduate education and research outputs in these areas. This situation raises several questions. For example, while the U.S. is ramping up funding for the medically-related life sciences, what are our competitors doing in terms of R&D investment strategies? In particular, what are those countries that are not only our economic but our military competitors as well, for example China and Russia, doing with their investments? In the following paragraphs, two areas of S&T will be used as examples—nanoscience and energetic materials—to look for answers to such questions.

Ernest H. Preeg, a Senior Fellow in trade and productivity at the Manufacturer's Alliance/MAPI, has written extensively about China's efforts to become an advanced technology "superstate." In a recent piece[28] in The Washington Times, he notes that in April 2005, Premier Wen Jiabao summarized the Chinese economic strategy in these stark terms:

> Science and Technology are the decisive factors in the competition of comprehensive economic strength...We must introduce and learn from the world's achievements in advanced science and technology, but what is most important is to base ourselves on independent innovation... [which] is the national strategy.

Examples cited by Preeg exemplifying China's advanced technology innovation are the planned launch of more than 100 satellites to form a global Earth observation system, the Dawning 4,000-A Shanghai supercomputer, and the Godson II central processing unit computing chip to support the 64-bit Linux operating system. Clearly, the Chinese are making strides in several areas of S&T by focusing their R&D investments. Stokes says that this year the Chinese hope to spend 1.5 percent of China's GDP on R&D, up from just 0.6 percent as recently as 1996. As a point of reference, in 2003 the U.S. spent 2.6 percent of its GDP on R&D.

Preeg has also written a recent book called The Emerging Chinese Advanced Technology Superstate,[29] which examines the rapidly evolving technological revolution in China. In it, he estimates that "even if the growth rate of Chinese R&D spending slows to only two-thirds of what it has been in the past, China will be spending 2.1 percent of its GDP on research by 2010. In dollar terms, this would amount to only about half of what the United States spends, but it would be 80 percent of the level of R&D spending in the European Union, and 40 percent higher than such spending in Japan." Preeg provides an abundance of statistics to emphasize the growth of the Chinese pool of scientific and engineering talent. According to his estimates, China has three researchers for every five in the U.S. This year alone, Chinese students will earn nearly 13,000 doctorates in S&E, about half the number granted in the entire U.S., and that gap is narrowing. By 2010, the number of Chinese S&E doctorates awarded is expected to exceed those earned in the U.S. Not only that, but the Chinese pipeline of young S&Es is full, unlike ours. In fact, S&E account for nearly three in five bachelor's degrees now conferred in China, compared with only one in three in the U.S. Table 3.1 summarizes a few of the comparisons between the two countries in areas of importance to high-technology competition and growth.

Time Frame	R&D Spending (in billions)		Researchers (in thousands)		S&E Doctorates		U.S. Patent Applications	
	U.S.	China	U.S.	China	U.S.	China	U.S.	China
1999	-	-	-	-	-	-	92,349	282
1999/ 2000	-	-	1,261	695	-	--	-	-
2001	-	-	-	-	25,509	7,601	-	-
2002	$277.1	$72.1	-	-	-	-	-	-
2005*	$330.9	$130.9	1,690	979	24,504	12,838	123,357	2,034

*Estimated

Table 3.1: High-Tech Comparisons Between the U.S. and China (Sources: OECD, Projections of Ernest Preeg, Manufacturers Alliance Cited in Stokes)

While it is certainly true that the quality of Chinese scientific talent is not yet up to levels found in the West, it is improving rapidly, and the output of that talent is growing rapidly in areas of special interest to their government. For example, Stokes observes that *"China is currently the second-largest producer of technical papers in nanoscience and nanotechnology. And more than half of Chinese research papers concentrate on chemistry, physics, and mathematics, the seed corn for innovation in advanced technology."* These fundamental areas form the pillars on which much of the research important to the DOD stands.

3.6   Nanotechnology Example

It is useful to take a closer look at the field popularly known as nanotechnology since it provides a good example of changing global trends and how they could impact on our future defense posture. In simple terms, nanotechnology involves using single atoms or molecules to make electronic circuits and devices, and is a key building block of the future. More specifically, it refers to the development and use of techniques to study physical phenomena and construct structures in the physical size range of 1-100 nanometers, as well as the incorporation of these structures into applications. It has been seen that China is rapidly becoming a major center for nanotechnology research. In fact, according to Stokes, *"China already leads the United States in some key nano areas. Chinese scientists can now produce carbon nanotubes 60 times faster than their American counterparts can. These tubes—actually cylindrical molecules of carbon—can form a substance stronger than steel and much, much more conductive than copper."*

Chinese technological strides in scientific areas of military importance, including nanotechnology, have already caught the attention of the DOD. To learn more about the Chinese effort, the ONR recently carried out an effort that used text mining techniques to examine the research outputs of a number of countries in nanotechnology to gain a better understanding of their S&T investment strategies. Text mining refers to the extraction of useful information from large volumes of text. In the ONR effort, Kostoff and other researchers performed an analysis[30] of the global open nanotechnology literature. The effort analyzed SCI records in order to provide the infrastructure of the global nanotechnology literature, e.g., the most prolific and most cited authors, journals, institutions, and countries, as well as the thematic (taxonomy) of the global nanotechnology literature from a science perspective. It also examined records from the Engineering Compendex[31] to provide a taxonomy from a technology perspective. Among their general findings were the following:

- Countries in the Far East have expanded nanotechnology publication output dramatically in the past decade.
- China ranks second to the USA (2004 results) in nanotechnology papers published in the SCI, and has increased its nanotechnology publication output by a factor of 21 in a decade.
- Of the six most prolific (measured by numbers of publications) countries in nanotechnology, the three from the Western group (U.S., Germany, France) have about eight percent more publications (for 2004) than the three from the Far Eastern group (China, Japan, South Korea).

- While most of the high nanotechnology publication-producing countries are also high nanotechnology patent producers of U.S. patents (as of 2003), China is a major exception, ranking 20th after the U.S. among the countries considered.

Table 3.2 shows the Kostoff et al. results in terms of the most prolific paper-producing countries for the year 2004. It can be seen that three countries dominate in terms of overall output of technical papers: U.S., China, and Japan. In terms of U.S. patents, the top three performers were U.S., Japan, and Germany. Note that China ranks last in the list of countries shown in terms of U.S. patents. Stokes suggests that one reason that China has not produced more U.S. patents is that they do a lot of research aimed at "reverse-engineering," so a lot of what they do is not patentable. There is also the possibility that some areas could be highly classified for military or commercial reasons.

Although detailed findings will not be presented here, a more thorough understanding of China's S&T investment in nanotechnology can also be gleaned from Kostoff's text mining efforts. It shows that the Chinese are focusing on several specific areas in the physical sciences. For example, its investment in scanning electron microscopy research is about twelve times that of the U.S. China is also heavily invested in research on nanorods and in Fourier-transform infrared spectroscopy.

The significant point that emerges from this kind of analysis is that the Chinese have carefully decided where to spend money in order to get the best return on investment, not only in terms of furthering their economic interests, but their national security interests as well.

Energetic Materials Example

Another area that can be used to analyze global trends in terms of their potential impact on DOD's efforts is the field of EMs. It is especially useful in that it provides a case study of several other points raised in this book, for example the dwindling size of the EMs S&E workforce and its technical output. Therefore, it offers a glimpse of where this field could be headed from a global perspective. The field of EMs is certainly one of major and enduring importance to the DOD, since they are a key component of many armament systems of crucial importance to maintaining U.S. military preeminence. Finally, it is one of the areas where most observers agree the DOD should be concentrating the efforts of its in-house laboratories and centers.

EMs is a term widely used to describe explosives, propellants, and pyrotechnics. In essence, they are mixtures of chemicals that undergo highly exothermic chemical reactions while converting condensed phase (solid or liquid) materials into rapidly expanding gases. The rate of this phase change is so rapid, and the volume of gas produced per unit mass of condensed phase material so large, as to be capable of producing destructive force. If the rate of reaction is supersonic (greater than the speed of sound), it is called an explosive. If the rate of reaction is subsonic, it is called a propellant. A pyrotechnic is a mixture of chemicals which when ignited is capable of reacting exothermically to produce light, heat, smoke, sound or gas.

Table 3.2: Most Prolific Countries (2004) in Terms of SCI Papers and U.S.P.O. Patents

Country	Number of Papers	Country	Number of Patents
USA	7512	USA	5228
Japan	4431	Japan	926
China	4417	Germany	684
Germany	3099	Canada	244
France	1900	France	183
South Korea	1592	South Korea	84
United Kingdom	1520	Netherlands	81
Russia	1293	United Kingdom	78
Italy	1015	Taiwan	77
India	830	Israel	68
Spain	727	Switzerland	56
Taiwan	706	Australia	53
Canada	690	Sweden	39
Poland	515	Italy	31
Switzerland	498	Belgium	28
Netherlands	492	Denmark	23
Brazil	455	Singapore	20
Sweden	435	Finland	17
Australia	434	Ireland	10
Singapore	372	Austria	8
Israel	347	China	8

The field of EMs is very old. In fact, it is believed they were first made by the Chinese more than one thousand years ago, when they discovered that mixtures of saltpeter and sulfur reacted rapidly, producing a highly-destructive force. Perhaps the earliest unequivocal reference to what would be today called gunpowder occurs in 1242 in the writings [32] of Roger Bacon, an English alchemist and monk. However, it was the Swedish chemist Alfred Nobel who gave much of the modern impetus to the field of EMs when he patented a composition called dynamite in 1867. It was one thousand times more powerful than black powder, and expedited the building of roads, tunnels, canals, and other projects worldwide. Because of the damage it could do, Nobel thought its invention would end all wars. However, as history has since shown, instead of ending war it ignited a global race to find newer, even more powerful explosives for military use.

Figure 3.7 depicts in greatly simplified form a few generations in the historical evolution of EMs, moving from the era of black powder to the very powerful plastic bonded explosives (PBXs) that came into use in many military applications following the Second World War. PBXs have since become the preferred choice for nearly all new naval and airborne weapons such as missile warheads. Despite past successes in the field of EMs, this area of scientific inquiry has not become static. Indeed, new areas of research continue to evolve, and the outlines of what might be called a 6th generation of EMs are already emerging. Examples include reactive materials (RMs) and nuclear isomers. RMs denotes a class of materials that typically combine two or more non-explosive solids which, upon ignition, react to release chemical energy in addition to the kinetic energy resulting when the high-speed projectile containing the reactive materials collide with the target. Nuclear isomers of the element hafnium are of interest as potential high-density storage media. Even more exotic is the idea of using antimatter for storing energy at extremely high densities.

The generations of EMs shown in Figure 3.7 resulted from years of sustained R&D carried out by a global workforce of researchers. The S&Es who carry out such work typically fall into three technical areas: synthesis chemistry, formulation and processing, and detonation science.

As the name suggests, *synthesis chemistry* focuses on the creation of new energetic molecules, and it is largely responsible for new breakthroughs in energetic materials. *Formulation and processing* involves making EMs in large amounts for use in test and evaluation experiments, and typically involves large-scale synthesis, formulation, pressing and casting. Finally, *detonation science* includes development of fundamental detonation and shockwave physics models; modeling of both detonation initiation and growth phenomena, and deflagration to detonation transition reactions in explosives and propellants; understanding mechanical and thermal initiation of explosives and pyrotechnics; design and development of initiation system components; and evaluating explosive and propellant sensitivity and hazard analysis. S&Es who work in detonation science are keenly interested in the behavior of energetic materials under all environments, and have contributed greatly to the weaponization of new materials, including ensuring the materials remain safe throughout their "stockpile-to-target" sequence.

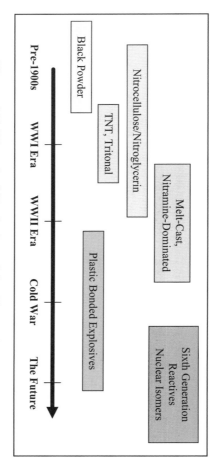

Figure 3.7: Historical Evolution of the Field of Energetics

Historically, the number of S&Es in the U.S. engaged in EMs R&D has been relatively small compared to the numbers in many other scientific fields. However, recent trends show that this already small number is shrinking further, a concern noted in several recent studies of the EMs field. Two of these studies,[33,34] provide a cogent summary of the current U.S. posture in EMs and are reviewed here in brief. The first is entitled *Advanced Energetic Materials*. It was carried out by the NRC through the Board on Manufacturing and Engineering Design, in response to concerns voiced by the Office of the Deputy Under Secretary of Defense for S&T and the Defense Threat Reduction Agency over the prioritization of scarce resources and issues related to maintaining and improving DOD's knowledge base in this critical defense technology area. The second study is called *National Security Assessment of the High Performance Explosives and Explosive Component Industries*. It was requested by NSWC's Indian Head Division to help address growing concerns over the ability of U.S. suppliers of high performance explosives (HPEs) and high performance explosive components (HPECs) to produce their products in the future. A HPEC is a weapon or subassembly of a weapon that utilizes HPEs as its source of destructive power, e.g., artillery shells, warheads for missiles, bombs, fuzes, detonators, etc. This study also considered the dwindling national investment in R&D which historically has led to the development of explosive materials for new applications. Because of falling DOD spending on munitions, much of the industry has faced reduced production orders and lower revenues bringing into question its very survival.

The study *Advanced Energetic Materials* found that, although all modern defense systems and weaponry rely heavily on EMs, the U.S. R&D effort has become small, fragmented and suboptimal, leaving this critical national technology area at risk. This suboptimal effort is characterized by severe resource limitations across the entire spectrum of EMs R&D. Significantly this situation arises just at a time when the current focus of the DOD is on limited theater actions that emphasize deployment of precision strike smart weapons that are smaller, cheaper, and at the same time more lethal against all target classes. All of these are demands that advanced EMs can and should address. And while the U.S. effort is shrinking, the study found strong indications that former Soviet states such as Russia are investing heavily in EMs R&D, and may well be exploiting technological breakthroughs that could have the potential to place U.S. armed forces at a substantial technological disadvantage in the future.

The study also called attention to the fact that funding for the nation's EMs technology effort is shrinking largely as the result of reduced defense spending on munitions R&D in recent years. Importantly, the study warned that *"Without the opportunity for the current workforce to train the next generation of expert scientists and engineers, much corporate knowledge may be lost. This knowledge is key to maintaining the current weapon stockpiles safely, to ensuring their performance, and to developing the next generation of energetic materials."*

As an indication of just how fragile the U.S. R&D effort in EMs has become, the authors noted:

The U.S. effort in the synthesis of energetic materials at present involves approximately 24 chemists, several of whom are approaching retirement. Few chemists are being trained to replace them. The committee considers these scientists to be a national resource whose productivity in terms of new energetic compounds has been very high. If the level of effort that these scientists have contributed is not fostered and maintained, the United States will lose the technological edge that it has gained as a result of their work. Attracting top synthetic chemistry talent to energetic materials research is possible only if the field is perceived to be scientifically exciting and financially stable.

The second study—*National Security Assessment of the High Performance Explosives and Explosive Component Industries*—was carried out by the Department of Commerce's Office of Strategic Industries and Economic Security, and reached similar conclusions. For example, it pointed out that the U.S. supplier base for HPEs has been operating under increasing stress since the late 1980s because of reduced production orders and lower revenues. It also called attention to the decline in spending on EMs R&D, noting that "*since its 20-year high in 1989, DOD spending on munitions research, development, testing, and evaluation (RDT&E) has fallen nearly 45 percent…[and that] according to current projections, RDT&E spending on munitions will plunge another 50 percent to about $820 million by 2005.*" The study found this issue to be even more serious in light of the fact that munitions RDT&E is also falling as a percentage of DOD's overall RDT&E budget. It was between 4 and 6 percent of the overall DOD RDT&E budget from 1986 to 2000, but then plunged to about 2.4 percent, a reduction that seemed likely to slow innovation and hinder the ability of the U.S. to field cutting-edge munitions technologies.

Finally, the study addressed the impact of these funding declines on workforce issues. Here its authors opined that "*Reduced RDT&E spending will almost certainly degrade the ability of firms and government organizations to hire and retain scientific and technical staff. Drastic budget cuts will send a loud signal to the chemistry and physics communities that there are few opportunities in the field of high performance explosives. Scientists and engineers will simply vote with their feet—opting to "follow the money" to financially healthier areas of research.*"

The impact of aging on the EMs S&E workforce will also take its toll, and replacing that workforce will not be quick or easy. According to NSWC Indian Head officials quoted in the study, it can take five years or longer to fully train a college graduate with a science or engineering degree to work with EMs. As the study notes, the long lead-time to train S&Es to work in this field, coupled with the anticipated retirements in the next 10 to 15 years, "*portends the development of a knowledge gap,*" in this critical defense technology area. It also raises important issues as regards the inter-generational transfer of corporate knowledge in this key area. The need for the DOD to address the issue of how it will preserve its existing corporate memory for use by future generations of defense

S&Es is a subject of great importance and is discussed in more detail in the next chapter.

Left unchecked, current workforce trends will likely result in serious employment issues with S&Es engaged in EMs as the current generation of workers retires over the next several years. As just one example, consider what has happened at the Weapon Division of the NAWC which is located in China Lake, California, once one of the DON's premier weapons laboratories. In 1985, it had a workforce of 10 to 12 employees engaged in active R&D of new energetic compounds. However, by 2003 that effort had dwindled to only two to three people carrying out applied development on a single material according to Robin Nissan, Head of the NAWC Chemistry and Materials Division.[35] Indeed, the number of synthesis chemists working in the entire Western world is quite small--fewer than 75 people by some estimates—and this entire effort is funded at the paltry rate of only about $10 million dollars/year.[36] While there are no comparable estimates for the numbers of S&Es working in formulation and processing and in the detonation sciences, it is likely they too are quite small and declining.

If the U.S. S&E workforce engaged in EMs R&D is shrinking in size, has this had an impact on its technical output? There is some evidence that it has. It can be found in the number of research papers published in technical journals. Kostoff and his colleagues have also investigated this technical area, again using text mining techniques. In this case they examined papers referenced in the following three data bases: the SCI, the Engineering Compendex, and the DTIC. As before, the focus in this investigation was on assessing the most prolific authors, journals, countries, etc. They utilized technical experts drawn from the various communities that comprise EMs R&D to compile a list of key words and phrases for use in querying the various data bases. Figure 3.8 displays results based on querying all technical papers in the SCI data base. It shows a significant decline in U.S. technical output in the field of EMs relative to that of the rest of the world.

The data also show that while China accounted for only 2 percent of total SCI-referenced papers in 1993-1996, its share more than tripled to 7 percent by 2001-2004. In terms of EMs publications only, China's share rose from 3 percent to 9 percent over this same period. It can be seen that Russia and Germany are also prolific in EMs publications.

Figure 3.9 shows the declining percentage of U.S. technical papers over time, both in terms of overall papers and those in the EMs field only, as compared to other countries. These results are based on citations in the SCI data base. It can be seen that in 1991, the U.S. accounted for about half of all SCI-referenced papers, a figure that dropped to just over 40 percent by 2004. The fall-off is more dramatic in the EMs field. Here

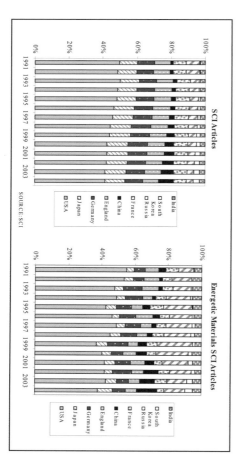

Figure 3.9: Publication Trends Showing Changes in Percentage of Articles Published by Country Overall and in EMs Only: 1991-2004

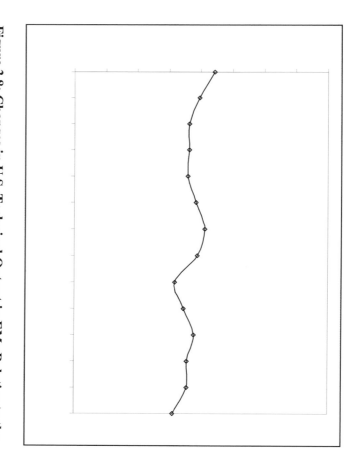

Figure 3.8: Changes in U.S. Technical Output in EMs Relative to the Rest of the World: 1991–2004

Erratum

Figure 3.8, p. 80

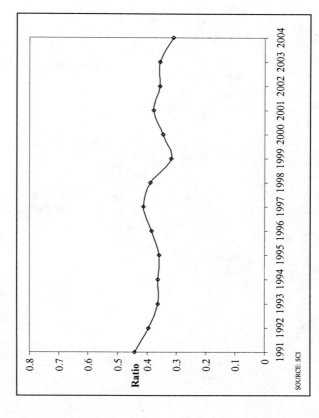

SOURCE: SCI

**Figure 3.8: Changes in U.S. Technical Output in EMs
Relative to the Rest of the World: 1991-2004**

the U.S. had nearly 55 percent of the SCI-referenced papers in 1991, but only about 36 percent in 2004. The slow but steady growth of Chinese papers over this same timeframe is also evident, both overall and in the field of EMs.

The Kostoff results do seem to indicate that the shrinking size of the U.S. S&E workforce engaged in EMs R&D has also resulted in a diminished technical output measured in terms of the number of papers published in SCI-referenced journals. These data do not, however, allow judgments to be drawn about the quality of this output or its impact in the field. These results do provide some insight into what two of our military rivals—China and Russia—are doing in the field of EMs. They demonstrate that Russia has maintained a vibrant effort in this important technical area since the early 1990s. In addition, they indicate that China is probably increasing its investment in EMs R&D aimed at developing new and very likely more powerful explosives for military applications.

The shrinking size of the U.S. S&E workforce engaged in various aspects of EMs R&D raises an important question: what can be done to preserve this critically important defense technology area? As will be discussed in the next chapter, one approach is to leverage DOD's in-house R&D expertise by collaborating with outside educational organizations. In fact, one such example involves an alliance called the Center for Energetic Concepts Development (CECD). Its purpose is to foster continued advancements in EMs S&T and manufacturing. Additionally, it will also help train the next generation of S&Es working in energetics through its graduate education and research programs. The CECD came about when NSWC's Indian Head division initiated a program to expand cooperation with universities and in turn increase the number of graduates with experience in EMs. The effort commenced in September 1998 with the signing of a contract between NSWC Indian Head and the University of Maryland. The agreement calls for the two organizations to work together to:

- Develop an internationally recognized energetics capability
- Develop the next generation of DON energetics experts
- Support DOD and non-military research priorities
- Access world-class experts in energetics and related disciplines
- Share experts and facilities.

Among other things, the agreement that established the CECD called for the University of Maryland to cost-share faculty time and to provide free graduate courses for NSWC Indian Head S&Es. More recently, the CECD has broadened its scope to include the energetics manufacturing S&T as well as modeling and simulation of energetic systems. It has also been funded to help train technicians at Maryland community colleges and through distance learning techniques. It is hoped that this alliance will eventually be expanded to include additional universities, national laboratories, and even private firms. Importantly, this kind of innovative arrangement could also serve as a repository of knowledge, helping to preserve for future generations vital scientific and engineering expertise that is currently dispersed around the country. This new organizational construct could also offer a model for other innovative arrangements involving both public and private sector partners that would help the DOD leverage its shrinking in-house S&T resources.

3.7   Trends, Data, and Modeling

In the introductory chapter of this book, reference was made to a joint Industry-Academia-Government workshop held in December 2004 just outside Washington, D.C. Called "National Security Workforce: Challenges and Solutions," it brought together key stakeholders in the U.S. national security enterprise to address pressing S&E workforce problems. At the workshop, attendees identified an urgent need for good models and the necessary data to feed them that could be used to provide accurate long-term forecasts of S&E workforce supplies and demand. They agreed that macro level data are probably sufficient for characterizing the current state of the national S&E workforce, but noted that these data currently lack specificity as regards needs attendant to the national security enterprise. To help rectify this situation, the attendees recommended several steps be taken including the following:

- Identify the kind of data needed to feed forecast models that affect workforce supply and demand (local, national, global).

- Assess its current quality and availability in order to determine gaps in the data, and institute steps to close them.

- Routinely collect data related to both supply and demand, including both global and national S&E supply estimates in the disciplines and sub-disciplines important to national security.

The workshop made one thing abundantly clear: there are currently no workforce forecasting models that incorporate the myriad variables that impact S&E workforce supply and demand. Moreover, even if there were good models, there are major gaps in the data that would be needed to populate them. Much of it is not collected today. Indeed, the rapid off-shore accumulation of technical talent makes it nearly impossible to develop accurate quantitative models of U.S. S&E supply and demand. Much of the data that would be needed is not and probably cannot be obtained, in part because of the dynamic nature of the global S&E enterprise and its growing interconnectedness. This enterprise has many loops and branches, and their effect is to introduce time-delayed effects and amplifications that are difficult to model in terms of their impact on our domestic supply of S&Es.

Because of this complex behavior, modeling this S&E enterprise can be considered as somewhat analogous to modeling and solving a non-linear control systems problem. Consider, for example, the possible effects that could result from reducing R&D funding available to U.S. universities by 5 percent in terms of impact on our global competitiveness. This reduction might well prompt a foreign-born student who would have otherwise come to the U.S. for graduate studies to join an educational institution in another country instead. There, he will add to their intellectual capital and the global competitiveness of that country. Concomitantly, his not coming to the U.S. represents a loss in so far as our ability to be more competitive on the world stage. Moreover, over time such a funding reduction might have a negative impact on the reputation of U.S. universities, making it more difficult for them to attract the best and brightest students. For example, post-9/11 visa and travel restrictions might send a signal to foreign students that the U.S. no longer welcomes them. This situation would then have another negative impact on our talent base and its technical output and impact and, ultimately, on our global competitiveness.

This very simple example illustrates that the decision branches introduced by the competing forces in the S&E enterprises can lead to time-delayed non-linear effects. Unfortunately, it is difficult to obtain a quantitative assessment of the exact degree of non-linearity introduced by such time-delayed loops. Nevertheless, it is possible to create conceptual models that are useful in thinking about the various factors that affect S&E supply and demand in the global enterprise, and how they could be impacted by time-delayed non-linear phenomena. By way of illustration, a few examples of such conceptual models are presented in the following sections. It should be emphasized, however, that these models are not unique—they are not data-driven. Other individuals attempting such models would doubtless pick and choose their own set of variables and relationships they consider most important, and would doubtless "wire" them together in ways different from those shown in the examples here.

As a first example, consider a so-called "base" model that follows the decision points that a U.S. born student encounters in deciding to enter into a S&E career in the U.S. This model also follows the path that a foreign-born student takes, either to enter into a S&E career in the U.S. or in another country. The education/career path of the U.S. student and the foreign-born student are essentially parallel processes that are nevertheless highly linked to each other.

The U.S. side of the model begins with a U.S. citizen of college age making the decision to attend a U.S. university. There are several factors that control this decision such as degree of high school preparation, perception of the worth of a college education, cost of attending, perception of the future job market, etc. The foreign-born student weighs similar considerations, including whether to attend in the U.S. or in another country. The decision process for the foreign-born student is, however, more complicated than for his U.S. counterpart. For example, the decision also involves uncertainties associated with living in a foreign country, potential language problems, less clearly defined opportunities upon graduation and, more recently, potential complications due to post-9/11 visa and travel restrictions.

At this point, there are two different pools of applicants: U.S. students who have decided to attend a university, and those foreign-born students who have also decided to study in the U.S. Now, these students must decide whether or not to pursue an S&E education path. Factors that influence this decision might include the perception of opportunities in S&E, the health of the U.S. economy, opportunities in other fields such as law, business or medicine, level of federal support of S&E, etc. The foreign-born student ponders similar factors and in addition, those related to being foreign-born, such as opportunities for remaining in the U.S. after graduation.

The next step involves the process of admittance into their desired educational program. Both sets of students must go through a competitive admissions process. The decision of what discipline/field to pursue also involves a decision tree. The foreign-born student has also to weigh the additional factor of being statistically less likely to be admitted than a U.S. applicant. Moreover, the issue of financial aid may well be more important to the foreign-born student than the one born in the U.S.

Following admission to a university of their choice and completion of all required studies, the students, both U.S. and foreign-born, become members of a pool of highly-trained S&E graduates who are now available to join the workforce. At this point, the two groups of graduates (U.S. and foreign-born) must decide whether or not to join the U.S. S&E enterprise. The foreign-born graduates also have to decide whether to try and remain in the U.S. or to take a position in the S&E workforce of another country, a very important decision for them. In fact, it is not only important to their future, but to the future strength of the U.S. S&E enterprise. As previously alluded to, this decision is one that is closely related to the issue of whether or not the U.S. will maintain its current preeminence in S&E.

Following the base model further shows that if the foreign-born graduate decides to remain in the U.S. and join its S&E workforce, then the two parallel tracks merge into one path that leads to the production of measurable technical outputs such as journal articles, patent applications, etc. Conversely, if the foreign-born graduate decides to leave the U.S. and join the S&E workforce of some other country, the base model paths remain divergent. In that case, the U.S. trained but foreign-born graduate will now be adding to the intellectual capital and technical output of that country as an economic competitor of the U.S. In fact, data trends discussed previously show that this is a growing trend that will continue to erode the current position of the U.S. as a global leader in S&E. As seen, many other countries, especially in Asia, are rapidly improving their educational systems and their production of S&E graduates and thus their technical output, and this is happening at what appears to be an accelerating rate as directly measured by numbers of papers, citation rates, patent counts, etc. Simply put, the stronger foreign competitors become, the more likely they are to attract and retain their own intellectual capital as well as that from other countries.

The base model captures a number of causes and effects, and therefore it can also be used as a perturbation model, making it useful as a tool for analyzing some of the principal concerns that face the U.S. S&E enterprise. In principle, such a model could be populated with representative data useful for relating input and output at each decision point, thus providing some level of quantitative estimates similar to those that can be obtained with some econometric models. The point here is neither to develop such a model nor to produce quantitative results, but rather to provide the reader with some suggestions as to how one might produce a model useful for making forecasts about the U.S. S&E workforce and the kinds of considerations such a model would have to address to give it credibility. It is also intended to highlight just how difficult it would be to construct an accurate model given the highly-dynamic nature of the global S&T enterprise and its S&E workforce.

Turning to the idea of the perturbation model, consider as an example the situation with regard to U.S. visa regulations, Figure 3.10. The effect of visa regulations and restrictions, including travel requirements, on the supply of foreign-born students applying to U.S. universities has been a highly discussed topic ever since the events of 9/11.[37] Many are concerned that intensified monitoring of foreign students will result in fewer students, especially graduate students, coming to study scientific and technical subjects in American colleges and universities and ultimately will reduce the supply of scientific and technical personnel available for employment in the U.S. Indeed, although these new restrictions were only intended to improve our homeland security posture, they can nevertheless have negative impacts. This very point was the focus of a recent piece[38] by William Wulf, who is president of the National Academy of Engineering. It suggests the U.S. may be trading the long-term health of its research and education for what he calls *"the appearance of short-term security."* More specifically, he has this comment about new U.S. visa restrictions:

Much has been written about the impact of new visa policies on students. Although the situation has improved somewhat in the last several months (as of this writing, the average time for processing visas for students is less than two weeks), I am still concerned because the distribution has a "long tail." Some students must still wait a year or more for visas, and some senior scholars, including a Nobel laureate, are still being subjected to lengthy, demeaning treatment. These cases, not the shorter processing time, are being reported in the international press, and as a result, instead of the United States being seen as a welcoming "land of opportunity," it is now seen as exactly the opposite. When coupled with new, demeaning procedures for photographing and fingerprinting visitors, we are not just discouraging students, international conferences in the United States, and collaboration with our international colleagues. We are dramatically altering the image of our country in the eyes of the rest of the world.

As Wulf notes, since many of these new rules and restrictions have only been in effect for a relatively short time, it is not clear how the impacts will play out, especially since there are many other factors, as already discussed, that affect the ultimate decisions of such students.

In addition to following the paths that either an American-born student or a foreign-born student takes in attaining a career in S&E, Figure 3.10 provides a conceptualization of how an increase in visa restrictions might ultimately impact the technical output of the U.S. S&E workforce. It can be seen that a reduction in the number of foreign-born students causes a concomitant reduction in the number of foreign-born graduates of U.S. universities, and consequently a diminution of the number of highly trained foreign-born S&Es available to join the U.S. workforce. This has two results. First, it produces a further reduction in U.S. technical output (e.g., patent and paper citation count). At the same time, it adds to the output of the country where the students decides to go—a net loss for the U.S. and a net gain for a potential U.S. economic competitor. Again, if a model such as this could be populated with appropriate data, it would show that a 5 percent reduction in the supply of foreign-born students for U.S. universities might have an even greater impact on the number of papers published by the faculty of those universities given the very productive nature of most foreign students. Such a causal model could also be utilized to study the effects of other influencing factors such as S&E workforce retirement rates, changes in the U.S. investment rate in R&D and so on.

Another significant consideration as regards the supply of American-born university students who want to study S&E fields has to do with the early education infrastructure in this country, K-12 for example. In this regard, consider the causal model depicted in Figure 3.11. It too shows the cyclical (feedback) effects that may occur if this infrastructure isn't improved in a way that produces more high school graduates who are interested and qualified to pursue S&E degrees at the college or university level.

It is worth noting that the timeframe covered in this model may be as much as twenty years: the time it takes a child entering kindergarten to ultimately obtain a Ph.D. degree in S&E and be ready to enter upon a research or teaching career. Unless the number of U.S. students that go on to careers in S&E can be increased, the effect will be to further decrease U.S. technical output. This, in turn, can cause additional high-tech American firms to off-shore their R&D to those countries that do have the requisite technical talent. Indeed, many prominent American corporate leaders—past and present—have commented on the connection between the state of America's early education infrastructure and its impact on U.S. innovation and productivity. Consider for example the recent congressional testimony[39] of Norm Augustine, former chairman and chief executive officer of Lockheed Martin Corporation, before the House Committee on Education and the Workforce, Subcommittee on 21st Century Competitiveness:

...with regard to seeking to recover from any ill-advised attempt to under-invest in research and education, it takes a very long time to produce additional productive research scientists. A youth wishing to become a mathematician, scientist or engineer must decide in ninth grade to take courses which preserve the option to pursue a career in any of these fields. This is a consequence of the hierarchical and interdependent character of a science or technology education. Further, the "leakage" rate in the process of producing credentialed researchers is very high indeed. In the field of mathematics, for example, based on current trends one must begin with 3,500 ninth-graders in 2005 to produce 300 freshmen qualified to pursue a degree in mathematics. Of these, about 10 will actually receive a bachelor's degree in the field. Finally, one PhD in mathematics will emerge in about 2019.

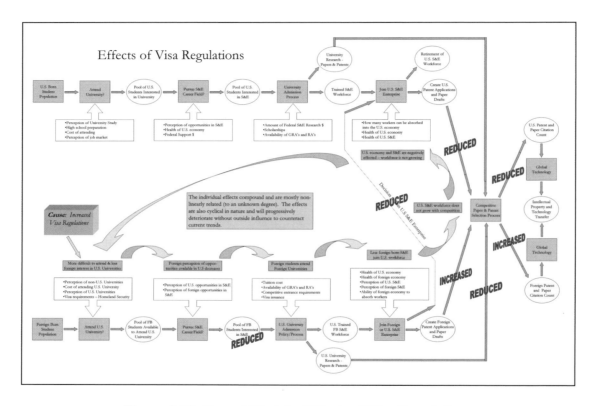

Figure 3.10: Causal Model – Effects of Visa Regulations
(See www.cecd.umd.edu for larger scale)

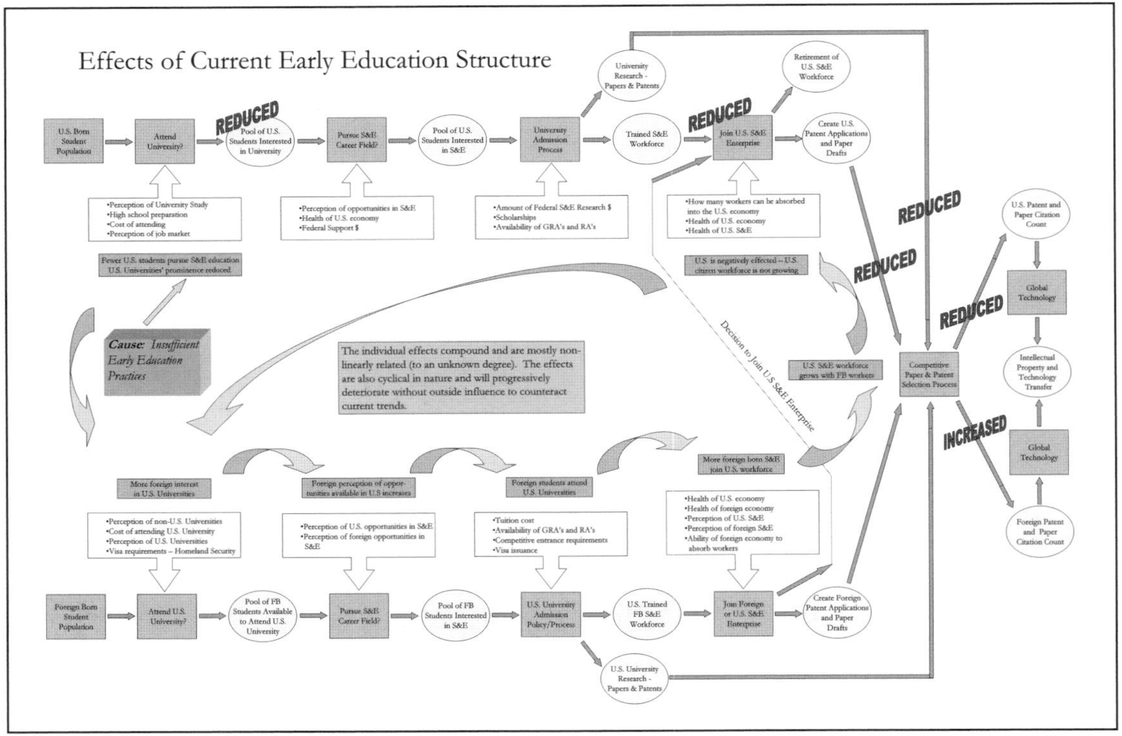

Figure 3.11: Causal Model – Effects of Early Education Structure
(See www.cecd.umd.edu for larger scale)

Returning to the subject of modeling U.S. S&E workforce trends, it can be seen that many of local, national, and global trends are interrelated so that a negative result in one can produce negative impacts in several others. It can also been seen that reversing some of the trends—like those characterizing the current U.S. K-12 educational infrastructure—is a very long term proposition, as previously noted. The value of models, even causal ones such as the examples presented here, is that they help us understand the trends and their implications for the future. It is very important to gain this understanding soon because of the long delays involved in building a workforce with the required skills to replace the S&Es of the Baby Boom generation, who are retiring just as the needs of national defense and homeland security are increasing.

The S&E workforce problem has many dimensions, and potential fixes vary from short-term to very long-term (many years). Only a few of the problems can and should be addressed by the DOD and by elements of the national security/homeland security enterprise. Indeed, some mention has already been made of efforts currently being undertaken by the DOD such as the SMART scholarship problem. The DOD and the Services also have many other efforts on-going or in the planning stage. Some of these are focused on getting more women and minorities interested in careers in STEM fields. As will be seen later in this book, the DON has been especially aggressive and innovative in its efforts to increase the pool of U.S. citizens qualified for S&E careers, including some aimed at the middle-school level. This is imperative, because the DOD and DON laboratory/center S&T workforces need revitalization now, not later. Moreover, because there is a direct causal connection between federal spending on S&T and its impact on research fields and their technical output, and on graduate training in those fields, it is essential for the DOD and DON to also reinvigorate their investment in S&T, especially that small portion that builds and supports the S&T workforces at the in-house laboratories and centers. This is in fact a major focus of the DON's N-STAR initiative (Naval Research—Science & Technology for America's Readiness). Because of its promise as a model for other similar initiatives, a detailed discussion of N-STAR and related initiatives will be provided in Chapter 5.

Chapter 4

Characteristics of the New Science and Technology Enterprise

4.1 Introduction

A sweeping reform effort aimed at transforming the way the Department equips, trains, and supports its military forces is underway in the DOD. RDT&E management processes are a part of this review, and a major aim of defense transformation is to foster the development and employment of joint military capabilities. The recent *Joint Defense Capabilities Study*,[1] chaired by Pete Aldridge, former Under Secretary of Defense for Acquisition, Technology and Logistics (USD (AT&L)), described the problem this way:

The Department's RDT&E resources and infrastructure are decentralized across the Components. In fast-moving technology areas, this decentralized approach to planning, programming, and execution results in inefficiencies, duplications, missed opportunities, and the inability to mass critical expertise in emerging areas.

Sustaining the dominance of military technology in the face of relentless change will not be easy. The rapid spread of new technologies now available worldwide raises the question of how the DOD can best identify and use them to create new military capability. The fact that these same capabilities are now readily accessible to our friends and foes alike makes it all the more important that we use them first.

Various change factors reveal that the DOD's current approach to managing its S&T enterprise will not produce the results its military must have for the 21st century. A recent white paper,[2] by the Institute for Defense Analyses calls for a *"fundamental re-examination and alteration"* of innovation practices, as global technological and economic changes have *"invalidated the very basis by which the DOD has achieved technological superiority. The global diffusion of capabilities in emerging technologies raises serious issues as to how those technologies with implications for advanced military capabilities can be identified, supported, accessed, and employed by the DOD to maintain DOD's position of technological superiority. Today's DOD S&T practices are largely unresponsive to this emergent environment."*

General John Jumper, who was until recently the Air Force Chief of Staff, characterized the need for technological advantage as follows: *"We are not interested in a fair fight.... We want to put overwhelming technology into the hands of our warriors."*[3] But as the IDA paper suggests, giving General Jumper the advantage he called for will require a fresh approach to the way DOD manages innovation. Finding the best approach must include examining the role of the laboratories and centers and the S&T workers they employ. This fresh approach must include global collaborations, a highly educated S&T workforce and its supporting structure, intergenerational transfer of knowledge, visionary leadership, and fully engaging military officers in S&T.

4.2   Collaboration and Networking

S&T breakthroughs today are increasingly the product of the collective efforts of networks of collaborators. These *innovation networks* are often global in nature, frequently draw on cross- and multi-disciplinary branches of science and engineering, and include players from both the public and private sectors.[4]

The growing need to network has inspired an upsurge of cooperative innovation agreements. These involve a diversity of participants, including commercial firms, universities, and research institutes to name only a few. Also, there are a growing number of government efforts in the U.S. to bring together public sector laboratories and centers with their private sector counterparts. The now familiar Cooperative Research and Development Agreement, is one example. In fact, a mix of university-government-industry linkages now characterizes many if not most technological sectors. Consequently, they are of increasing importance to the new DOD S&T enterprise.

Strategic alliances, joint ventures, and intimate supplier-producer relationships are proliferating across the globe. Agreements encompass a wide range of activities, such as *"joint ventures, research corporations (e.g., research pacts, joint development agreements), technology* exchange agreements (e.g., technology sharing, cross-licensing, mutual second-sourcing), direct investment, minority/cross-holding, customer-supplier and customer-user relationships, R&D contracts, one-directional technology flow agreements (e.g., licensing, second-sourcing), manufacturing agreements, marketing agreements, or services agreements."[5] Some groups, often referred to as generation networks, focus on generating new technology. Their activities include "licensing and cross-licensing agreements, technology exchanges, visitation and research participation, personnel exchanges, joint development, and research consortia."[6]

Technological globalization is of mounting concern to defense policy makers, and has prompted a DSB study called *Technology Capabilities of Non-DOD Providers*.[7] The task force that conducted the study argued that, as a globalized technology base has come to dominate development of capabilities in a number of critical areas, it has grown much faster than DOD's own funded research efforts. As a result, the Department increasingly depends on it. The task force recommended that DOD *concentrate the efforts of its limited laboratory and center resources on those unique military technologies and systems which are of crucial importance to maintaining US military preeminence.* Examples cited include armament systems, undersea warfare, surveillance, identification, targeting systems, and nuclear systems.

It should be noted that the study occurred during the still heady days of the "Dot-com" bubble, when it appeared private sector R&D investments were increasing almost exponentially. Those days are past, and, as noted in Chapter 2, commercial investments in basic research are

Sources for DOD Technology Innovation		
	DOD Sources	**Non-DOD Sources**
Non-DOD Unique S&T	DOD should not fund (use non-DOD sources)	DOD should use commercially developed S&T
DOD Unique S&T	DOD should fund (but no longer attracts "best and brightest")	DOD should utilize, but must change its procurement

Table 4.1: Sources for DOD Technology Innovation

dwindling in many, if not most, areas. Even so, it is true that DOD's laboratories and centers cannot be the technological leaders in all the fields of importance to national defense. Therefore, they must have a sufficient number of qualified S&Es who can perform as peers, in all areas of potential interest to national security, in the worldwide networks that generate new knowledge. Otherwise, DOD risks falling behind in the global race for knowledge. The DSB task force recommended the paradigm for DOD innovation depicted in Table 4.1.

Collaborating and networking now more frequently than ever draw on cross- and multi-disciplinary branches of science and engineering for scientific breakthroughs. A good discussion of this trend is in a recent report by the Council on Competitiveness, which notes:

Historically, advances in knowledge came through the efforts of individual investigators with specific disciplinary specialties.... Today, however, innovation tends to occur more frequently at the intersection of disciplines and, indeed, sometimes drives the creation of entirely new ones, such as nanobiology, network science or bioinformatics.

Advances in medical technologies integrate biology with physics, mathematics, materials sciences and software engineering. Innovation in the IT sector is built on research that spans a range of sciences...and increasingly, social sciences and the unique dynamics of particular industries, as IT planning becomes integral with business and organizational strategy.

At issue is not a choice between a single discipline specialization and multi-disciplinary research. The ability to innovate at the intersection of disciplines, by definition, implies the need for strong disciplinary expertise. But, knowledge silos simply won't drive innovation in a much more interconnected world. Indeed, they will inhibit it.

The changing nature of innovation demands new knowledge and learning networks that can facilitate communications and collaboration at frontiers of many disciplines and that can cross organizational boundaries between academia, industry and government.

While academia has been exploring interdisciplinary approaches for decades...such efforts at universities remain insufficient –

and have yet to emerge as a core focus of the national research enterprise.[8]

Similarly, advances in areas such as medical technologies are underpinned by biology, physics, mathematics, and materials sciences. This is equally true of the fields important to the DOD. Clearly, "working in the seams" between various disciplines is becoming progressively more important not only to U.S. innovation generally, but also to DOD's effort to maintain its technological lead over its adversaries.

Since at least the advent of the GWOT, it has also become clear that other, non-technical fields of study are also important to national security. Examples include foreign languages, cultural and religious studies, and history. John Holzrichter, an assistant to the director of the DOE's Lawrence Livermore National Laboratory and president of the Fannie and John Hertz Foundation, addressed this point in a paper for *Physics Today* on attracting and retaining R&D talent for defense.[9]

Whatever the future brings, national security requires more highly talented and motivated experts...not only in the traditional S&T disciplines, but also in biology, computer science, and other fields. Equally important are technical experts who can work...with nontechnical experts in social sciences such as diplomacy, policy-making, political science, behavior, economics, and international law, to name a few. Talented people, trained to deal with new knowledge and unknown conditions, are needed to respond to large numbers of unexpected—and sometimes "should have been expected"—situations.

4.3 A Highly Educated S&T Workforce

As discussed above, the unprecedented nature and extent of technological change today is often the result of interdisciplinary research. Such research "*integrates information, data, techniques, tools, perspectives, concepts, and/or theories from two or more scientific disciplines or bodies of specialized knowledge to advance fundamental understanding or to solve problems whose solutions are beyond the scope of a single discipline or area of research practice.*"[10] Indeed, interdisciplinary thinking is now an integral feature of most scientific inquiry as the result of four powerful drivers:[11]

- The inherent complexity of nature and society
- The desire to explore problems and questions not confined to a single discipline
- The need to solve social problems
- The power of new technologies

This new context means that those planning S&T careers will have to be more highly trained, and their education will have to be much more broadly-based and multidisciplinary.

Educators and professional societies are already seeking new teaching methods and curricula that can produce S&Es who can thrive in such a dynamic and globalized environment. This is especially true for engineering, with its rapid off-shoring of many types of jobs previously done here at home. Such changes have raised many questions about the future of professional engineering in the U.S., including whether it will even have a future.

These concerns prompted a major study at the National Academy of Engineering (NAE) to assess the next generation of engineering in the U.S.[12] Because of the near impossibility of predicting the future, the study participants helped bound the analysis by using a scenario-based approach that utilized alternative visions.

In a Phase I report, the Academy concluded that if the engineering profession is to define its own future, it must:

- *Agree on an exciting vision.*
- *Transform engineering education to help achieve the vision.*
- *Build a clear image of the new roles for engineers, including that of broad-based technology leaders who in the mind of the public and prospective students can replenish and improve the talent base of an aging workforce.*
- *Find ways to focus the energies of the different disciplines of engineering toward common goals.*

Importantly, "*if the United States is to maintain its economic leadership and be able to sustain its share of high-technology jobs, it must prepare for a new wave of change.*" To do this, the engineering profession must "*educate the next generation of students so as to arm them with the tools needed for the world as it will be, not as it is today.*"

A Phase II NAE assessment focused on the education appropriate to the engineer of 2020: "*The economy in which we will work will be strongly influenced by the global marketplace for engineering services, evidenced by the outsourcing of engineering jobs, a growing need for interdisciplinary and systems-based approaches, demands for new paradigms of customization, and an increasingly international talent pool.*" Therefore, undergraduate education "*needs to be reshaped to attract students to the profession, prepare them to compete in a global marketplace, and ensure that America's pre-eminence in engineering is not lost.*"[13]

Recognition of these new needs has led engineering organizations to recommend changes in degree programs. As the NAE states, "*it is evident that the exploding body of science and engineering knowledge can not be accommodated within the context of the traditional four year baccalaureate degree.*" For one, academic institutions should "*introduce interdisciplinary learning—today typically reserved for graduate programs—into the undergraduate engineering curriculum.*" Further, those with bachelor's degrees should be viewed as "*engineers in training,*" and the master's should be considered the engineering "*professional degree.*" Similarly, the Board of the American Society of Civil Engineers has adopted a policy that "*supports the concept of the Master's degree or equivalent as a prerequisite for licensure and the practice of civil engineering at the professional level.*"[14]

In sum, the NAE studies and other reports point out a number of implications for engineering and related fields. One, the M.S. degree will likely become a minimum requirement for entering most fields of engineering as a professional. Two, and perhaps more important, both undergraduate and graduate training will have to be much more diverse in content than it is today, involving not only fields such as mathematics and science, but also exposure to the humanities and training in analytical, communication, and foreign-language skills. In addition, because "*the half-life of cutting-edge technical knowledge today is of the order of a few years,*" colleges and universities will have to offer advanced technical training to working engineers, who will need this continuing education to maintain their technical relevance.[15] The same factors changing the engineering professions are also affecting the sciences and mathematics. There too, success will require more advanced education, exposure to various disciplines, and continuous training.

Ph.D.-Level Education or Equivalent

These evolving requirements have important implications for the DOD laboratories' and centers' S&T workforce. A Ph.D. or equivalent should be the goal for each journeymen member, especially any who aspire to cutting-edge work. Further, because many of the evolving fields of D&I will likely involve knowledge and skills not typically acquired through purely academic training, significant job experience and training that come through working with others who know the field will become crucial. While the goal should be a S&T workforce comprised wholly of S&Es with an education to the Ph.D. level, including post-doctoral work in many instances, the reality is that this will not be possible for all new hires. In those cases, a vigorous effort must be made to help ensure that those workers hired with B.S. or M.S. degrees receive the equivalent of a Ph.D. level educational experience.

Continuing educational opportunities that provide sufficient credits for a Ph.D. or its training equivalent are therefore increasingly necessary. Opportunities could be provided via on-site training or through distance learning methods. The subject of post-baccalaureate certificate programs has been widely discussed in the literature, and the Council of Graduate Schools has assembled a number of best practices in this area.[16]

One example is a certificate-based graduate engineering program that partners Virginia Polytechnic Institute and State University (Virginia Tech) with NSWC's Dahlgren Division. This program supports retention and professional development of technical personnel. It is considered unique in that it centers on graduate certificates that represent a concentration of expertise in a technical or scientific skill area. By linking coursework with practical research problems at NSWC, the program fosters focused skill building and improved workplace problem-solving capabilities. If they desire, participants can combine certain certificates with a dissertation or thesis to obtain a Virginia Tech graduate engineering degree.

The University of Maryland's CECD, which has already been discussed in Chapter 3, is another example of how cooperative efforts between universities and DON laboratories and centers can provide continuing educational opportunities for in-house S&Es.[17] As noted, this alliance partners the university with the NSWC's Indian Head, Maryland Division to foster advancements in energetics manufacturing, science, and research, while also helping to train the next generation of energetics S&Es through its graduate programs on campus and via distance learning.

Efforts such as these should be more widely exploited for a number of reasons. Distance education, for example, would allow universities to partner with others in the DON to build scientific and technical competence at globally distributed naval installations, while also furthering their own academic missions. Also, cooperative efforts offer the additional benefit of a collaborative, networked laboratory/university context.

DOD S&T Academy

As mentioned above, several DSB reports used membership in the National Academies of Science and Engineering as a metric of the quality of research organizations. Since universities dominate academy membership, the DSB concluded they are the prime performers of both basic research and engineering. It further judged profit-making companies, also well represented on the membership rolls, as strong players in engineering and technology development, along with various FFRDCs, national laboratories, UARCs, and non-DOD government laboratories. Importantly, because the DOD laboratories and centers are not well represented in academy membership, the DSB rated them to be of lower quality, and therefore unable to attract the best S&Es.

Their many scientific breakthroughs and engineering feats, plus the demographics of the academies' members and criteria for membership, suggest that Academy membership is an inappropriate quality metric for the DOD laboratories and centers. For example, only members can nominate other members, so who you know can be as important as what you know or even what you have done. Membership may also be awarded for other than technical achievement. For example, several members are retired Navy flag officers selected primarily because of their leadership and management skills. Consider the following: RADM Albert R. Marschall, elected in 1990 *"for outstanding management, stressing highly professional leadership in all phases of vital large-scale worldwide facilities programs"*; RADM Millard S. Firebaugh, elected in 2000 *"for innovation and U.S. Navy leadership in submarine design, propulsion, and construction"*; and RADM Robert Wertheim, elected in 1977 for *"contributions to national strategic programs, particularly engineering leadership of ballistic missile systems."*[18]

Academy membership is highly esteemed regardless of its inadequacies as a measure of quality, and yet the DOD laboratories and centers will most likely never have significant numbers of S&Es on the rolls. For one, membership is relatively small compared to the total population of eligible S&Es (both academies have just over 2,000 members each). Much more important is that a great deal of the work done for DOD is classified and often not even publicly attributable to the individuals who carry it out. Therefore, gaining public recognition for such work—an important academy consideration—is often impossible.

Given these circumstances, there is a real need for some institutional mechanism within the DOD that can provide the recognition its top S&Es deserve for their work, including classified work. One possibility is some kind of S&T academy, perhaps analogous to the national academies, membership in which would be reserved for those whose achievements on behalf of national defense are outstanding. Our national leadership would acknowledge election to membership, and would award prizes and/or privileges accordingly. For example, members might be awarded a stipend to work on a project of their own choosing, or to support something like a sabbatical for work with colleagues in the private sector.

Recognizing the contributions of the DOD S&T personnel to national security is not only appropriate—it can also attract technical talent. Acknowledgment, especially from peers, of the importance of contributions to national defense could provide a magnet for attracting some of the nation's best talent to work at the in-house laboratories and centers. In a major study of the Department's human resources, the DSB commented *"National leaders, at every level, need to speak to the American public, on an on-going basis, about the value of public service—both civilian and military."* Therefore, the Secretary of Defense *"should charge the Service Secretaries as a group with the responsibility to develop, execute, and fund an outreach strategy."* Such efforts *"should be a critical component of the Department's human resources responsibilities."*[19] An S&T academy would be a productive way for the DOD leadership to verify the value it places on research and engineering and on the men and women who carry it out.

Hiring, Retention, and Motivation

Human resource management systems must be tailored to enhance not only personnel recruitment, retention, and motivation. The current longevity-based CSS clearly will not suffice, and the jury is out as to whether the follow-on NSPS will prove up to the task. In fact, Congress recently exempted most DOD laboratory and center personnel from its operation until 2008, when the NSPS must prove it is up to the task. A letter from several members of the Senate Armed Services Committee to Secretary of Defense Rumsfeld explains that the new system will actually provide *less* flexibility for the laboratories and centers. The Congress therefore wants assurances that this will not happen:

> *Section 1101 of the National Defense Authorization Act for Fiscal Year 2004, chapter 99, section 9902(c) ...allows for the inclusion of the laboratories in NSPS after October 1, 2008, but only if the Secretary determines that the flexibilities provided by NSPS are greater than those already provided.*[20]

Whether the one-size-fits-all NSPS will in fact have both the flexibility and agility to serve changing personnel needs will not be known until the implementation details are finally decided, which could take several years. In the meantime, the DOD has in essence frozen further flexibilities for the current demonstration personnel systems. This threatens to hamstring the laboratories and centers at a critical juncture—just when they are trying to adapt to the new requirements of the on-going process of defense transformation.

Today, the DOD laboratories and centers confront a classic Catch-22: consistently doing first-rate work requires the ability to hire top scientific and engineering talent, yet they have difficulty hiring such talent because of the widespread—and unfounded—perception that they no longer do first-rate work.

A growing chorus of negative assertions has amplified this perception. Consider the following examples from recent DSB studies. In March 2000, the DSB declared the laboratories and centers *"are not competitive with leading industrial and university laboratories in terms of innovation and technical leadership."*[21] Another study from June of the same year asserted *"DOD laboratory directors are unable to obtain or retain the services of not only the 'best and brightest' scientists and engineers but even those of average capability."*[22] Both studies use little quantitative evidence to support those claims, for which there is little quantitative basis. Rather, these are likely the views of the authors, who are principally drawn from the private sector, which often considers itself a competitor with the in-house laboratories and centers.

Also, corporate marketeers and various interest groups, incentivized to lobby for the commercial sector, often promote such statements. A good example is the pro-defense industry lobby called Business Executives for National Security (BENS), which aims, among other things, to privatize as much defense work as possible.[23] Such groups often proclaim without reservation—and also without substantiation—that the private sector can do the work not only better, but also cheaper and faster as well. Further, many officials of various kinds, including elected officials, are interested in shifting government work to the private sector. They denigrate Federal employees and employment with blanket declarations—again unsubstantiated—that the private sector can perform the work more efficiently than civil servants.

A recent study by a tri-Service NRAC panel expressed concern over how these negative views damage DOD recruitment and retention efforts. The panel observed that indeed *"Perceptions are hurting the ability of the labs to attract and retain required talent."*[24] The authors called on the DDR&E, who oversees the DOD's S&T enterprise, to secure a commitment from the Secretary of Defense and the Service Secretaries to reaffirm the need for, and importance and value of, their laboratories.

John Holzrichter, mentioned above, also addressed the impact of this perception on hiring and retention at the DOE national laboratories. In his experienced view:

There are many ways to fix the twin problems of recruiting and retaining bright, talented, creative people for defense-related R&D. One of the most effective ways is for our national leadership to more clearly acknowledge the importance of workers' contributions in the defense sector, and in particular the importance of R&D. National security R&D is one of the best collective investments our nation has made. [25]

Holzrichter is saying is that yes, there are perception problems, but the nation's leaders can and should neutralize these by compellingly addressing the importance of government service to the country's security. Although he does not specifically call for a DOD S&T academy, his research lends support for creating one.

In addition to perceptions, the laboratories' and centers' declining capacity to offer appealing incentives to prospective talent is hurting hiring and retention. Incentives include not only tangible items, such as compensation and benefits, but also intangible ones, such as important and challenging work, access to state of the art facilities and equipment, adequate and stable funding to pursue important ideas, high-quality professional colleagues, and pride in the employing institution.

The problem is that one-size-fits-all personnel systems, such as the CSS, designed for the entire defense workforce, cannot offer these kinds of incentives. In fact, for the laboratories and centers currently operating under congressionally authorized personnel demonstration systems, the challenge for NSPS is to offer flexibility. Regardless, as discussed later in this chapter, attracting and retaining top-level talent requires human capital management systems tailored to scientific and engineering career paths. These systems must also permit the selective and appropriate rewarding of a few key individuals—the top performers.

Other factors have shackled hiring and retention at the laboratories and centers in recent years (most of these were discussed in more detail in Chapter 2). Five cycles of BRAC, arbitrary personnel cuts, intermittent hiring freezes, high-grade controls, forced contracting out of work, and budget cuts have curtailed investments in research facilities, state of the art equipment, and employee development and training. The authority of laboratory directors has been eroded, as services once under their local control have been turned over to various centrally or regionally managed and often geographically remote organizations, supposedly to save money. Such services include public works, base operations and support, finance and accounting, and human resources. As just one illustration, employees at the NAWC aircraft division in Patuxent River, Maryland now depend on a human resources office in Philadelphia to process their personnel transactions.

Although an unintended consequence, this centralization has in particular impaired hiring and retention efforts. Not surprisingly, it has diminished system responsiveness and slowed the delivery of services to laboratory and center employees. As just one example, regionalization of personnel services in the DON has significantly increased the time it takes to fill jobs, which all too often means high-quality S&Es accept other employment offers in the interim.[26] It has also impeded the timely processing of financial transactions—it often takes weeks to pay new hires. Finally, regionalization of base operations and support in the DON under a newly-established centralized command threatens to curtail investment in new facilities and high-technology equipment even more than before, a point elaborated on below.

In sum, the growing inability of the laboratories and centers to offer prospective S&Es—especially the best and brightest—sufficient incentives to hire and retain them is a serious impediment to the DOD's S&T effort. A shrinking pipeline of available new talent exacerbates the problem. This shortage is especially acute in certain areas of engineering and the physical sciences of particular importance to the DOD. The graying of the overall workforce, accelerating as the Baby Boom generation retires, aggravates the problem even further, and adds an additional element discussed later in this chapter: what should be done about inter-generational transfer of knowledge.

Important and Challenging Work

In the past, the DOD laboratories and centers have sought to offset pay disadvantages relative to the private sector by offering benefits such as the opportunity to do important and challenging technical work. Such work has long been recognized as a major attraction to high-quality employees. As the 1962 Bell report (see Chapter 2) noted, *"Having significant and challenging work to do is the most important element in establishing a successful research and development organization."* One reaction to the concern Bell expressed was the creation of the discretionary ILIR program.

The 1991 Adolph Commission report also noted the importance of interesting work for morale among laboratory staff,[27] while a more recent DON report put it this way:

One of the strongest attractions offered by the Naval in-house S&T enterprise is the wide variety of interesting, challenging and important work assignments, free from the pressure to introduce a new product every year, which is a pervasive pressure in most industrial laboratories today. This is important because fundamental research often does not produce a mature product or process for 20 years or more.[28]

Once again, the laboratories and centers are being stripped of the ability to perform a role just as that role is becoming increasingly important. In other words, at a time when challenging work is particularly vital to attract a new generation of the best technical talent, it is becoming more difficult to provide it. Two reasons for this, as noted, are the significant decline in discretionary money available to laboratory and center directors, and the long-term trend toward outsourcing of work to the private sector.

The impact of this outsourcing on internal technical competence has been discussed, but there is another, perhaps even more important, impact: outsourcing also means the remaining S&Es are increasingly saddled with overseeing the work of contractors. Monitoring contracts does not provide the same satisfaction as actual technical work, and therefore acts as a disincentive to remain in the Government. It also makes the higher pay in the private sector even more attractive than it might otherwise be.

4.4   S&T Support Structure

In early 2000 the SECNAV enlisted the National Academy of Public Administration (NAPA) to study potential DON human resource management systems. The results appeared in *Civilian Workforce 2020: Strategies for Modernizing Human Resource Management in the Department of the Navy*.[29] It described the current CSS as "ponderous, bureaucratic, slow, and unresponsive," and noted that "it is not a great leap of logic to predict that organizations with people-management systems that are controlled by over 2000 pages of law and regulations, with literally years required to make change, will not be among the winners in the war for talent." Indeed, the current system presents "an almost insurmountable barrier to achieving flexibility and agility in responding to new requirements. Change requires years of persistent effort with uncertain or negligible results." Consequently, the "flexibility to succeed in the 2020 environment is not possible within the constraints of the current federal civil service system. The inherent Title 5 concept *of 'one size fits all' will not serve the needs of the Navy nor the needs of the diverse technical and support communities upon which it depends."*

As the NAPA makes clear, the one-size-fits-all approach to human capital management codified in Title 5 of the United States Code will soon be totally ineffective. One key element to the burdensome regulations is the CSS compensation system, which has the following problems: narrow pay ranges with cumbersome processes for moving to another range; little relationship to the realities of labor market dynamics for key scientific and technical occupations; and limited ability to reward excellence.

The Academy recommended a new human capital management system tailored to today's scientific and engineering career paths. This system should maintain pay comparability at the 50th percentile for key occupations, with the ability to pay more for highly qualified personnel. Other basic features of a system should include a process for market-based pay within the parameters of a general schedule or pay bands, authority to pay a limited number of senior scientific and technical employees above pay caps, and elimination of civil service job protections for poor performers.

Other study groups have reached similar conclusions, and in fact encouraged the Secretary of Defense simply to utilize an existing authority to create a personnel system specifically for the laboratories and centers. The NRAC, for example, in its tri-Service study of DOD's corporate research laboratories, argued as follows:

The Panel recommends that the Secretary of Defense fully utilize the authority granted him by Section 1114 of the National Defense Authorization Act of FY 01, and any other authorities granted by Congress, to establish a separate personnel system for the scientists and engineers in the Services' corporate research laboratories.[30]

The Section 1114 authority granted by Congress is indeed powerful, because it took away from the Office of Personnel Management the authority to manage DOD's S&T laboratory personnel demonstration projects and vested it in the Secretary of Defense. This authority in essence gives the Secretary the tools to set up tailored personnel systems.

Studies of the Air Force laboratories have also focused on Section 1114. One example is the NRC's *Effectiveness of Air Force Science and Technology Program Changes*. After examining in great detail the problems the Air Force faces in hiring and retaining top-notch scientific and engineering talent, the authors commented:

A key step toward alleviating this situation would be for Section 1114 of the FY 2001 National Defense Authorization Act (Public Law [P.L.] 106-398) (U.S. Congress, 2000) to be implemented. Because this issue transcends the Air Force, such direction would have to apply to all of the service laboratories and would therefore be effective only if directed by the Secretary of the Defense.

Therefore, the authors recommended: *"The Secretary of Defense should immediately direct the implementation of the provisions of Section 1114...so that Department of Defense laboratory directors have the ability to shape their workforces."*

Competitive Compensation: The Essential Element of a Quality Organization

Although not the only major factor, compensation is often determinative in employee job decisions, and the field of compensation is moving in new directions.[31] Howard Risher, who has served as a Senior Fellow in the Wharton School's Center for Human Resources, has studied compensation extensively, including its use in government organizations. In a review of the best thinking in 41 major companies with large R&D centers, he discussed compensation models that foster improvements in both organizational and individual performance.[32] Basic features of this concept include:

- Broad-banding to replace the traditional salary structure
- Competency-based pay to shift the focus from the job to the individual
- Increased emphasis on market alignment rather than internal equity
- Expanded role of variable pay linked to group/team performance
- Increased emphasis on recognizing and rewarding individual achievement

Salary decisions based on the individual's value to the organization, rather than on defined job duties or relative job value, encourage the kind of employee development particularly important in fast moving, dynamic areas of science and technology.

An essential ingredient for an effective compensation structure is pay bands. *"One of the key changes,"* Risher identifies among the companies he studied is *"the shift from the tightly structured ranges and centralized control of traditional programs to salary or career bands. The bands increase flexibility to respond to labor market trends and to recognize individual growth and contribution.*" Further, "*salary increases within the band are then linked to assessed competence, with the larger increases granted to the individuals who demonstrate new or enhanced competence.*" Career bands involve separate broadband structures for defined job families. They typically "float" with the market for the job family. Risher also notes the expanded role of cash incentives, including group incentives. In sum, salaries are aligned with individual competence, while cash incentives are used to reward results.

Another aspect of the new compensation model involves "market alignment." Recognizing they have to offer whatever it takes to attract and retain the best talent, the most competitive organizations have shifted away from the traditional goal of internal equity, so familiar in Government organizations, to external competitiveness. Many high-technology companies now set pay levels for primary job families solely on the basis of market data. As noted in the Introduction, many defense and aerospace companies are now having to pay significantly higher levels of compensation to hire cleared or clearable S&Es in fields such as aeronautical and electrical engineering, where salaries are soaring by 10 to 15 percent annually.

The differences between the new compensation models of successful R&D companies and those of the Federal government are telling. Companies pay valuable individuals much more than others if the market dictates. Conversely, Federal pay approaches, including those incorporated in the S&T personnel demonstrations used at a number of DOD's laboratories and centers, lump all S&Es into a single career path. Further, they pay roughly the same salaries to those situated at the same place along this path. Most importantly, salaries are not based on true pay comparability dictated by market alignment, a bias that leads to egalitarian compensation approaches not conducive to rewarding particularly valuable individuals. This focus on the "average" misses a lesson competitive organizations have learned. As Peter Drucker states, "*The averages serve the purposes of the insurance company, but they are meaningless, indeed misleading, for personnel management decisions.*"[33]

The point is that highly competitive organizations understand that success depends on having a lot of good people and a few extraordinarily good people. How many of the latter an organization needs is subject to debate, and depends on a number of factors. For an organization primarily engaged in S&T, the figure is probably around 10 percent of the S&E workforce. Dr. Hans Mark, former DDR&E, once remarked, "*The presence of a few individuals of exceptional talent has, to a very large degree, been responsible for the success (and even the existence) of outstanding research and technology development organizations.*"[34] Similarly, a former president of the National Academy of Sciences put it this way: *"In the advancement of science, the best is vastly more important than the next best."*[35]

In fact, plenty of evidence indicates that just a few highly talented individuals in a research workforce produce much of its output. One study for example observed that *"scientific output is concentrated amongst relatively few scientists,"*[36] while another concluded that the top 10 to 15 percent of scientists contribute about half the papers published, and that this is consistent across a range of fields.[37] Work comparing the distribution of citations to the distribution of publications,[38] also shows that the inequality in the former is much greater than in the latter. This suggests that fewer than 10 percent of scientists are responsible for about half of citations in a field.

Whatever the exact number, it is clear that for any scientific organization to be world class, it must have a small cadre of outstanding employees. It must therefore have the flexibility to offer whatever it takes to attract and retain them. Needless to say, such individuals are in increasing demand around the world, and will only work for organizations that provide them the benefits—tangible and intangible—they desire.

4.5 Career Management of the S&T Workforce

Tailoring human resource management systems means simply recognizing that different *communities* of employees have different career path needs, and that there are different ways among those paths to measure success. For example, a scientific and engineering career path follows a certain set of core competencies, common knowledge, skills, and experience, which differ in many ways from those of a lawyer, contracts specialist, or accountant. Career management in each of these communities should consider things like a common core training curriculum, common culture and professional identity, identifiable career paths, and links to community related professional associations.

In fact, community management is quite familiar within military circles, and its benefits have long been understood. Within the DON, for example, Surface Warfare Officers and Engineering Duty Officers (and various others) chart and control the career paths and other aspects of these communities. The assumption is that utilizing some aspects of community management for the civilian population would, as it does for the military, provide for a national and unified understanding of the health of the community and the needs of its members. It would also provide the refreshment and retention strategies that will work best for each specific workforce.

In this regard, measuring health is one of the most important aspects of the overall career management approach. Here, the health of the community's workforce is defined by reference to a set of metrics considered important to the workforce. Examples of such metrics might include education levels, age, diversity, experience, and hiring/attrition rates. They could also include measures of the size and qualifications of the potential applicant pool.

The DON has in fact attempted to implement some form of community management. As noted in the Introduction, in response to the NAPA report, *Civilian Workforce 2020: Strategies for Modernizing Human Resource Management in the Department of the Navy*, the VCNO chartered a task force to examine civilian personnel issues. The resulting study recommended that the DON utilize community management. It identified 20 potential communities: science and engineering, facilities, information technology/management, environment, logistics, contracts, human resources, legal, financial, education and training, administration, analysts, medical, security and law enforcement, intelligence, media and public relations, community support, acquisition program management, manufacturing and production, and wage grade (included as a distinct community). It also envisioned that a leader for each community would help coordinate such efforts as:

- Promoting the needs of the workforce to top DON leadership
- Providing a forum for Navy S&Es to discuss issues of importance, including their relationship with S&Es in the private sector
- Fostering professional development by understanding and sharing best practices and advocating for training and development budgets
- Providing professional recognition for community members
- Developing templates for various career tracks

The DON S&E community, with almost 21,000 members, was considered diverse enough that it would need various career tracks from which workforce members might choose. Examples included systems engineering, R&D, R&D technical expert, and engineering management. Moreover, DON planners also deemed S&T a distinct subset of the S&E community, because of its own unique characteristics. This would include both S&T performers and managers at the NRL, ONR headquarters, and within the warfare/systems centers.

Despite initial efforts to implement this approach—including an overall plan for a Civilian Community Management Division (CCMD) within the Office of the Deputy Chief of Naval Operations for Manpower and Personnel (N1)—little has come of the effort. This issue should be revisited and strongly supported. The point is not to provide a status report on the DON CCMD effort, but to demonstrate the value in championing S&E careers within a community context. This would recognize the special needs of that workforce and allow for tracking its overall health with an eye to rapidly developing and deploying new strategies to address emerging requirements.

4.6   Stable Funding

As long as analysts have been studying the Federal laboratory system, they have emphasized the connection between effectiveness and stable, adequate funding. In 1945 Vannevar Bush, in his now classic book, *Science: The Endless Frontier*, helped establish the Federal Government's approach to its laboratories and centers. He argued, as the title suggests, for the need for continued, intensive support for scientific research.[39] In 1983 the highly influential Packard Commission recommended multiyear funding of R&D, noting, *"the direction and performance of Federal laboratories is less than optimal because of serious problems with the continuity of research funding. Supporting high quality research requires stability and a long-range view."*[40] One year later, a White House Science Council study addressing the Packard panel's recommendations argued that funding instabilities impede the planning and operation of research programs. It also suggested that Congress adopt a biennial budget for Federal laboratories.[41]

Stable funding is especially important in S&T, where basic research projects can extend for several years before being terminated or transitioned. One report observed that *"the extraordinary historical productivity of several of America's most innovative and important research laboratories (Bell Telephone, DuPont, Xerox, IBM) is often traced to their commitment to a long-term, stable support to researchers and programs with productive pasts and promising futures."*[42] The DOE white paper discussed in Chapter 2 also identified funding instability caused by budget fluctuations as a barrier to strategic workforce planning in the national laboratories. In effect, these fluctuations resulted in erratic hiring patterns and contribute to the difficulty in hiring the technical staff needed to meet mission requirements.

System Agility

Today, system agility may be one of the biggest impediments to hiring and retention in the DOD laboratories and centers. The *Quiet Crisis*,[43] a report prepared for the CNR, introduces this subject with an insightful quote from an acknowledged expert in corporate management, Peter Drucker:

What people mean by bureaucracy, and rightly condemn, is a management that has come to misconceive of itself as an end and the institution as a means…. The hospital does not exist for the sake of the doctors and nurses, but for the patients…. The school does not exist for the sake of the teachers, but for the students.[44]

Indeed, some analysts have estimated that navigating administrative mazes consumes as much as half of the working time of S&Es.[45] As officials in the DOD and Services increasingly sought to monitor and control in-house work, they constructed increasingly complex administrative structures of review. The Task Force 97 report, in a statement echoed countless times since, lamented that S&Es in the DOD laboratories were *"buried within a wearisome administrative structure."*[46] The Bell Committee report recommended eliminating layers of management;[47] a 1962 DSB study chartered by Dr. Harold Brown, a former DDR&E, bemoaned inappropriate bureaucratic regulations;[48] in 1970, a blue ribbon panel appointed by Secretary of Defense Melvin Laird reported that excessive centralization in the OSD had impaired effective civilian control of the laboratories and centers, and that overly large staffs caused unwarranted paperwork, delay, duplication, and expense;[49] a study by Booz, Allen & Hamilton of the DON R&D process found that increased centralization directly increased reviewing authorities and their staffs, which in turn stifled initiative, stultified progress, and diluted resources.[50]

Ironically, centralization wrested many crucial support functions from the laboratories and centers just as Vice President Gore's highly touted National Performance Review advertised decentralization as a means to increase government competence. Although centralization's advocates promised increased efficiency at lower cost, to date, the results contradict the claims, largely because efforts were designed to optimize the efficiency of the service-providing organizations, not the customers. Indeed, these new support organizations are just the kind of bureaucracies Drucker describes, because they operate for the most part as an end to themselves.

The bottom line is that flexibility—not rigid central control—is essential to an effective RDT&E organization. This is especially true in regard to financial resources and other forms of necessary support. Centralization of operation and support functions, and the erosion of the laboratory director authority resulting from the centralization of various services, operate counter to these needs. The Adolph Commission worried about such trends, noting "*an effective laboratory has sufficient local operating authority to execute [its] responsibilities in a rational, effective manner. Laboratory management must have authority to plan, organize, staff, and direct its technical program as well as all necessary support services to ensure that the technical program is not impeded by inadequate support. The support services should be organic to the laboratory.*"

Similarly, as the *Quiet Crisis* points out, "*the effects of this excessive control can be delays in facility and equipment procurement that, in turn, delay R&D projects, some of which are critical to national security requirements, or lengthen the time it takes to process personnel actions, thereby exacerbating the difficulties in recruiting high-quality S&Es. These bureaucratic constraints can threaten work quality and employee morale to the point where a talented researcher concludes that the system is unworkable, and he or she departs for employment in industry or academe.*"

Bureaucratic regulations have always presented difficulties, but recent changes are crippling the capacity for dealing with them. In the past, most of the laboratories and centers found work-arounds. However, as their overhead and discretionary resources continue to decline, their ability to "make the system work" in the face of growing bureaucratic constraints is rapidly decreasing. As the system fails to deliver quality services to the employees when needed, frustration will increase, and more and more will seek employment elsewhere.

In sum, many DOD and Service centralization initiatives have been driven almost completely by the desire to cut costs, or to run things as cheaply as possible. Almost no attention has been given to the impact on the delivery of services, which in turn affects the direct missions of the laboratories and centers. Nor has any attention been paid to the impact on their recruitment and retention efforts.

Facilities and Scientific Equipment

Various studies have demonstrated the necessity of up to date facilities for effective laboratories. The Adolph Commission's report to Congress noted that state of the art equipment and specialized laboratory facilities "*appropriate to advancing the leading edge of relevant technologies are necessary to fully exploit the creative potential of scientists and engineers.*" Further, "*new technical facilities must become available at the rate for which technology advancement is desired; there is a direct cause-and-effect relationship.*" Similarly, a 1987 DSB task force commented, "*R&D depends upon the use of state-of-the-art equipment and facilities. Providing [these] is made difficult by rapidly changing technology which results in equipment becoming quickly outmoded and by the increasing cost of renewing such equipment and facilities.*"[51] This task force also argued, "*without very good facilities and equipment, in some areas, even an excellent researcher cannot compete with a mediocre researcher who does have the facilities.*"

Unfortunately, for a variety of reasons, it is becoming almost impossible for many DOD laboratories and centers to invest as required to do cutting-edge technical work. For one thing, they must have the authority to do so when and where needed. However, construction of most new facilities is subject to the idiosyncrasies of the MILCON process. Politics, competing priorities, and other factors can delay construction projects for years. Indeed, few facilities—especially those needed for general-purpose RDT&E—make it through the process these days. Without MILCON funding, many laboratories and centers are left with the unattractive alternative of upgrading old facilities, even though renovation is often less efficient and much more expensive than new construction. The simple fact is that MILCON is often the most cost effective approach.

Getting approval for general-purpose research buildings has always been difficult because, in the decision-making process for construction dollars, RDT&E does not compete well with readiness-related activities. In the DON, decisions for RDT&E MILCON priorities are largely under the control of the operational side of the Navy via the Deputy Chief of Naval Operations for Fleet Readiness and Logistics (N4). On October 1, 2003, the CNO stood up a new organization within N4 called Commander, Naval Installations (CNI) to centrally manage all installations, including laboratories and centers.[52] However, the principal mission for CNI—indeed, its *raison d'etre*—is to enhance fleet readiness, a goal that therefore receives much higher priority than long-term RDT&E facility needs. It is therefore important that a clear path be established to ensure the availability of adequate funding for scientific equipment and related infrastructure.

4.7 Inter-Generational Transfer of Knowledge

The evidence clearly indicates that the DOD can expect a critical problem in the near future, how to transfer the corporate memory inherent in their workforces to the next generation of S&Es.

What is corporate memory? Most individuals in an organization, especially high performers, are storehouses of specialized knowledge. This includes not only subject matter expertise, but also familiarity with less tangible matters like organizational folklore, culture, and oral tradition, all of which facilitate smooth and efficient performance. Such collective wisdom further consists of experience in specific projects, networks with clients and contacts, and awareness of an organization's informal relationships and decision-making processes. All together, this learning represents corporate memory. It needs to be identified, safeguarded, shared, and passed to the next generation of workers, as it is one of an organization's most valuable—yet often unrecognized and under appreciated—assets. Corporate memory is especially important to high technology enterprises involved with R&D.

Any time employees leave, whether voluntarily or involuntarily, they take with them some of this knowledge and lore. When the separation is voluntary, there may be an opportunity to pass along some knowledge to a successor. However, even this opportunity is dwindling today in many in-house fields of expertise because they are already one-deep, and this has a real impact on the apprentice programs that have served the defense laboratories and centers so well in the past. When, however, the separation is involuntary—as has frequently been the case with the forced downsizing in the DOD laboratories and centers for the past 15 years—there is a loss of corporate memory.[53]

Innovation in R&D organizations has been widely studied, and the results state unambiguously that much of it depends on informal networks that enable vital collaboration. One study identifies three critical variables that underpin the value-added creation process: skilled human assets, skilled senior leadership, and adequate resources. If any one of the three is absent, value-added creation is unlikely. Competitive organizations use a strong culture to bond these three variables in ways that cultivate core competencies and capabilities.[54] The important point here is that loss of employees, in particular key individuals, disrupts informal networks and collaboration processes, destroying the corporate memory needed to prosper.

The impending retirements, along with the dwindling pool of available new workers, will have a tremendous impact on corporate memory. The Center for Innovation Management Studies (CIMS), which has examined inter-generational knowledge transfer in the corporate context, notes that in the short term the *"Xers and the upcoming mini-boomers are not going to feed the innovation engine that the U.S. industry needs to stay competitive."* The problem is even worse *"because we haven't been educating them in science and technology."* As a result, *"the crisis looms as those boomer-aged professionals, especially technical professionals, begin to retire."*[55] A few companies are already addressing the knowledge loss that goes with retirement. Some, for example, are rethinking their policies on mandatory retirement, looking at such alternatives as phased retirement, rehiring retirees as consultants, and creating mentor programs.

The crucial point about mentoring programs is that the need must be recognized and the process formally structured to accomplish their goals. Such programs involve experienced employees teaching newer ones things that make up the company's corporate knowledge. This function could be full-time or simply one task of the employee's job. The *"final reality,"* according to CIMS, is that *"organizations must prepare for this predicted decline in the number of technically competent workers."*

In his book *Lost Knowledge*, David DeLong has offered a number of prescriptions for dealing with this pending loss of corporate memory. [56] One of the examples he documents in some detail is that of NASA. By 2006, at least 50 percent of the administration's employees will be eligible for retirement. According to DeLong, NASA still hasn't come up with a complete replacement program. Moreover, even if NASA could find enough younger S&Es to replace the ones retiring (or the ones lost already), it cannot replace or even substitute for the experience and knowledge they will take with them. Indeed, DeLong observes that the problem is so severe NASA no longer knows how to send a manned spacecraft to the moon. The S&Es who performed so well at that in the 1960s and 1970s have already left. The important knowledge they have taken with them includes not just that of the technical and engineering requirements, but of the procedures and processes involved in preparing an organization for such an undertaking. Regrettably, this tacit knowledge was not well documented, and that part of NASA's corporate memory is gone forever. Reconstituting it, if it can be done at all, will be expensive and time consuming. In essence, taxpayers will be paying again for something they already paid dearly for years ago.

Importantly, the Commission highlighted the significance of corporate memory and mentoring in training. *"Over the last half century,"* it observed, *"the nuclear weapons complex has successfully managed the training of such individuals primarily through mentoring and on-the-job training. In some critical skill areas this training takes about five or more years to gain sufficient experience to make design decisions."* The Commission noted Sandia's new intern program to supplement on the job training, and Los Alamos' effort called "TITANS," or Theoretical Institute for Thermonuclear and Nuclear Studies." This is a course, primarily focused on nuclear weapons design for new personnel, *"to facilitate knowledge transfer of archived data and cross training."*

The need to invest in inter-generational transfer of knowledge is also underscored in an article in the *Harvard Business Review*. The piece focuses on both the explicit and tacit knowledge an organization's best employees have stored in their heads. This knowledge—*Deep Smarts*—cannot be transferred on a *"series of PowerPoint slides or downloaded into a data repository."* Instead, it can only *"be passed on in person—slowly, patiently, and systematically."* Further, these Deep Smarts are experience based, and *"can't be produced overnight or imported readily, as many companies pursuing a new strategy have discovered to their dismay. But with the right techniques, this sort of knowledge can be taught—if a company is willing to invest."* This is precisely the kind of acumen NASA relied on to put a man on the moon and the DOE used to provide the U.S. with state of the art nuclear weapons throughout the Cold War.[57]

The demographic trends threatening the ability of the DOD laboratories and centers to preserve their Deep Smarts in areas vital to national security also imperil a growing number of private defense and aerospace companies. The NRC has noted how these trends jeopardize the inter-generational transfer of specialized technical knowledge and skills: *"The change in the age-experience composition of the work force occasioned by the decrease in defense spending raises serious questions about mentoring and the generational passing on of knowledge in the industry."* As a result, *"Many companies are now planning formally arranged mentoring procedures, as well as more training programs for new workers. When new engineers are hired, mentoring and teaming arrangements have to be carefully planned to capture the experience of those about to retire. Nevertheless, experience is lost, as is efficiency, when work tasks involve significant learning curves."*[58]

In sum, the imminent retirement of large numbers of S&Es in the DOD, the resulting loss of corporate memory, and the lack of programs designed specifically to ensure the inter-generational transfer of this memory all have potentially dismal consequences. The DOD faces the prospect of having to reconstitute a number of its important research areas unless steps are taken soon, and in many of these areas there is too little private sector interest to allow a commercial alternative. An example is EMs research, or studies of the basic science associated with reactive materials such as explosives and propellants. At present, there are still enough potential S&E mentors in the overall population to facilitate inter-generational transfer of knowledge, but their numbers are dwindling every year.

4.8   Civilian Leadership

Since the publication of the Bell report in 1962, there have been some 100 major studies on improving the ability of the DOD laboratories and centers to execute their missions. Remarkably, little progress has occurred with many of the issues those reports have often raised. There are many reasons for this inaction. However, one important reason has received little attention: the lack of consistent, unqualified, high-level support and advocacy for reform efforts from both the military and civilian leadership.

The NRAC wrote one of the few studies to call attention to this problem.[59] Its look at the corporate laboratories the Services operate considered among other things why reform efforts had so often foundered in the past. Among reasons they identified were several so-called "practical impediments."

For example, because of the bureaucratic tendency to resist change, senior leadership at all levels must make an unequivocal commitment to endorse exceptional reforms, and insist subordinates implement them. That is, for efforts to succeed decision-makers at all levels must make reform a high-priority issue, and take the time to understand the issues and the solutions proposed. And they must demonstrate a continuing commitment to implement their decisions.

For some time, analysts have been concerned over such growing problems with the S&T leadership in the Federal Government. For example, in 1992 a major report from the National Academies noted, *"As scientific and technological knowledge continues to expand at a rapid rate, the government needs ever greater capacity to formulate, carry out, and monitor S&T policies and programs and their effects. The need for highly competent and dedicated scientists, engineers, and other experts in top policy and program management positions in the federal government has never been greater.*"[60] The authors worried that the Federal Government was facing increasing difficulty in recruiting talented and qualified individuals. Causes for this difficulty included:

- Stringent and confusing post-government employment restrictions
- Long, burdensome, and intrusive nomination and Senate confirmation processes
- Strict and costly conflict of interest provisions
- Overly detailed requirements for public financial disclosure
- Pay uncompetitive with comparable positions in the private and nonprofit sectors
- The high cost of living in Washington, D.C.
- Increased public scrutiny of private life

- Decreased capacity of government to execute effective programs

- Lower public esteem for and prestige of public service

The report notes that these factors apply to all PAS positions government-wide, but can "have a differential impact on the government's ability to attract researchers from academia and industry and managers with technical backgrounds from industry," because "Government service does not usually further the careers of practicing scientists and engineers." It also points out that it took the first Bush administration an average of nine months to fill key S&T positions, up from six months in the Reagan administration. "This lag in filling positions," the authors assert, "has a significant and harmful effect on the government's ability to manage ongoing programs and to undertake S&T-based initiatives."

A good example of how such factors can hamstring qualified candidates with bi-partisan Senate support is that of Gordon England, the former Secretary of the Navy whom President George W. Bush nominated for the position of Deputy Secretary of Defense. His confirmation was delayed for several months while members of the Senate Armed Services Committee struggled to overcome a rule "requiring the Pentagon's top officials with pensions from private companies to buy an insurance policy that locks the value of their benefits while in office."[61] It seems MetLife, the only company that has provided such insurance, was no longer willing to do so. Although this impediment was eventually overcome, it prevented England from quickly taking over a critical position within the DOD.

RAND's National Defense Research Institute (NDRI) carried out a more recent study of this problem. Requested by a DSB Task Force on Human Resources Strategy chartered in 1998 by the Under Secretary of Defense for Acquisition and Technology, the NDRI report, *Department of Defense Political Appointees: Positions and Processes*, was published in 2001.[62] The DSB task force incorporated some of the findings into its report released in February 2000.[63] The NDRI's detailed analysis found that since the creation of the DOD in 1947, the number of PAS positions (including the Military departments) grew from just 12 to 45, as of May 1999. As the number grew, the percentage of available time those positions were filled declined. For example, during the Truman administration, PAS positions were filled an average of 98 percent of the available time. During the Clinton administration, the average was 80 percent, and in fact, since the Carter administration, seldom have all positions been filled.

A study published earlier this year reached similar conclusions about how tenure issues related to PAS positions suffocate reform. Called *High-Performance Government: Structures, Leadership, Incentives*, it noted that in the late nineties the executive branch experienced vacancy rates in appointed positions that frequently exceeded 25 percent. As a consequence, *"if it takes the better part of a year to complete a presidential-appointment, it follows that many departments and agencies will need to limp along. 'Acting officials are disinclined to initiate anything, they shy away from difficult administration problems, they avoid congressional testimony and public representation for most of their decisions.'"* And so we have a growth of inaction.[64]

As briefly mentioned just above, one difficulty in sustaining any reform initiative stems from the organizational placement of the Office of the DDR&E a placement that hampers its power to provide centralized leadership and advocacy for the S&T enterprise. Recently, the Congress has become concerned about this matter. In its report on the fiscal year 2005 National Defense Authorization Act, the Senate Armed Services Committee included a section (911) entitled *Study of roles and authorities of the Director of Defense Research and Engineering*. This called for a DSB study on the requirements *"to enable the DDRE to effectively perform the required mission."*[65]

The Committee expressed concern about the strength of the DDR&E, but did little other than direct a study of the situation by the DSB, a step unlikely to do much to alter the *status quo*. If so, the ingredients needed to provide stable leadership and top-level advocacy for the DOD's S&T enterprise will remain largely unavailable just as the GWOT and other factors require strong, hands-on leadership.

Visionary Leadership

Studies have also accentuated the importance of leadership, and identified the salient characteristics of effective laboratory and center managers. The Packard Report noted, *"The quality of management is crucial to a laboratory's performance. Federal agencies must insist on highly competent laboratory directors."*[66] The Adolph Commission also identified *"an inspired, empowered, highly qualified leadership committed to technical excellence through support for excellence, creativity, and high-risk/high-payoff initiatives,"* as an essential element of a world-class laboratory. The Commission specified several attributes of a good technical director:

- High standards of qualification in technical background and technology management experience

- Commitment to a creative work environment where individual initiative is encouraged and nurtured
- A long-term management perspective of planning, accomplishment, and resource commitment
- Willingness to undertake developments recognized as being high risk, and having high payoff potential

In other words, successful RDT&E organizations have leaders with vision who balance long-range challenges against immediate technology needs and risk against payoff. Moreover, such leaders are adept at choosing the areas in which to work, divesting work that has become more appropriate for other performers (transition), judging scientific and technical merit, and orchestrating the conditions that foster innovation.

Because of the long-term horizon for much scientific and technical effort, leadership continuity is also important (not only at the field level, but at the headquarters level as well, a point addressed in more depth below). Again, many study groups have acknowledged this need, such as the Task Force 97 report:

> In virtually all the Defense Laboratories authority and responsibility are divided between a commander and a technical director. Often, the commander is not technically qualified for the position [director of technical programs]. Even in the cases in which the commander is technically qualified, he falls victim of a military career management policy which requires his rapid rotation without regard for research program requirements. [67]

Rotation of commanding officers has a ripple effect, affecting both continuity and corporate memory. A 1976 review of Navy R&D management commented:

> To some extent, rotation of line officers into RDT&E activities was considered desirable. First, it was necessitated by the career requirements of naval officers. Second, it brought in-fleet operating experience and understanding of new weapons and systems developments back to the laboratories. Nevertheless, some technical directors felt strongly that rotation compromised the stability and continuity of the technical program and inhibited the internal management and control of the activity. [68]

Related to the matter of management continuity is the larger issue of military versus civilian management of the laboratories and centers. The DSB has commented:

In previous examinations of the in-house laboratories, the problem of military versus civilian leadership has been considered critical.

It is generally conceded that competent management...requires a sound knowledge of the military problems encountered in actual field and combat situations. This has been the leading argument for maintaining military management control of the Defense Laboratories.

Nevertheless, in a carefully planned program, it is not out of the question to have civilian personnel who are thoroughly versed in military affairs from a quite practical viewpoint. It is as possible for civilians to understand the military environment as it is for military personnel to be trained in technical areas.[69]

The DON sought to balance this joint relationship through the dual executive—having both a Commander and TD. It codified the approach in a Secretary of the Navy Instruction (SECNAVINST 3900.13A of November 1, 1963). This instruction clarified that the military officer commanding the laboratory was responsible for overall management and the usual functions of commanding, such as ensuring compliance with legal and regulatory requirements, serving as a liaison with other military activities, and generally supervising the quality, timeliness, and effectiveness of the technical work and support services. On the other hand, he was to delegate line authority and assign responsibility to the TD for the technical program, including directing its planning, conduct, and staffing; evaluating the competence of personnel; serving as a liaison with the scientific community; selecting subordinate technical personnel; exchanging technical information; and also ensuring the effectiveness of the program.

Dual executive management helped minimize the problems of discontinuity associated with officer rotation, and generally worked well for over 30 years until the Navy Secretariat cancelled it in 1995.[70] Since then, the position of director of the warfare center field divisions has continued to weaken. In fact, it has been abolished in the NSWC and NUWC. Another blow has been the creation of the CNI, which also has eaten away at the authority of the laboratory/center commander by turning control of most base operations support functions over to CNI regional commanders.

All this adds yet another element to the major recent changes in the DON warfare/systems centers, and intensifies the challenge of providing the visionary leadership they need in this new century. Certainly the DOE national laboratories have acknowledged the growing necessity for world-class leadership, appointing only internationally recognized S&Es to their laboratory director positions.

4.9   Military Technology Officers

In the past, the DOD laboratories and centers have benefited from staffing young military officers with technical backgrounds alongside civilian scientists and engineers. The civilians learned something of war operations while the officers gained appreciation of how S&T develops future military capability. The eminent scientist and engineer Theodore Von Karman, the first head of the Jet Propulsion Laboratory at the California Institute of Technology and a man with close connections to the Air Force throughout his career, described the benefits of the synergy this way: *"Scientific results cannot be used effectively by soldiers who have no understanding of them, and scientists cannot produce results useful for warfare without an understanding of the operations."*[71]

Of the three military Services, the Air Force has utilized this principle to the greatest extent. In fact, staffs at Air Force laboratories and centers have, in the past, included significant numbers of officers with technical backgrounds. Von Karman himself argued that 20 percent of the scientists and engineers in its S&T organizations should be military officers.

Recently, the technical staffs of the laboratories and centers in all three Services have experienced dwindling numbers of military officers with technical degrees. This trend was noted with concern in the recent NRAC tri-Service study cited earlier.[72] The panel unanimously agreed on the importance of uniformed S&Es and worried that the Services were not fully exploiting that capability. Although this issue was only briefly investigated because it was not part of the formal study charter, the panel did find the following:

- The number of officers with advanced degrees in S&E is declining, especially in the Army and Navy.

- In the past, such officers usually spent one or two tours in Service laboratories/centers.

- Tours provided an understanding of the RDT&E process and problems.

- This helped link the laboratories/centers and the operating forces.

- The need for officers may become even more important, given the emphasis on providing our military with overwhelming technological superiority.

The panel asserted that the Services could, to their considerable long-range benefit, improve development, support, and promotion of uniformed S&E personnel. More specifically, they argued that a tour in a laboratory or center should be considered as one of several possible career-enhancing (rather than the opposite) assignments available to a junior officer, no matter what his or her background and specialty. The rationale is that it is as important to understand the development of war-fighting technologies as it is to understand the intricacies of DOD's Planning, Programming and Budgeting System process, which is commonly taught to officers as part of their acquisition training.

A comprehensive study of the Air Force's S&T workforce for the 21st century cogently articulated the benefits of officer tours in laboratories and centers:

> [The ideal military lab] has military officers flowing from the operational and acquisition communities into the lab for one or more tours and then back into those communities. This constant leaving serves to ingrain the lab with the military thinking and...the military communities with the best of the innovation from the laboratories. This...explicitly recognizes that the best technology transition occurs through the transfer of ideas and minds rather than reports and devices. A further benefit... is the presence of a number of general officers who are cognizant of the capabilities of the S&T laboratory. Since [they] serve as the senior leadership base for a military organization, the recognition of value in the S&T laboratory depends in part on the ability of the general officers to understand the contributions and culture of the laboratory.[73]

This report reaffirmed the 20 percent figure Von Karman argued for, noting such a strong military presence benefits a "high tech" Service. It further observed that this percentage had fallen to 11 percent, and recommended it be increased to 15 percent as an interim step. Moreover, it argued this increase should "occur in the context of a career management plan for military S&E officers that has them serving in the lab as well as in the operational commands and the acquisition community," because one goal "should be to have more General Officers cognizant of this critical part of the Air Force."

A more recent study of the Air Force by its Scientific Advisory Board (SAB) expressed even greater alarm over the military S&E presence in their laboratories, calling the situation a "graveyard spiral."[74] The SAB found worsening trends in accession and retention as well as in development and management. 1999 data showing the inventory of military S&Es, the current authorization, and the numbers needed for self-sustainment suggest:

After the first 4 years… the inventory roughly matches the sustainment or authorized curves. However… a crisis in military engineers will soon occur, if it has not occurred already. The Air Force has been accessing during those 4 years at slightly over half the rate necessary to sustain the force. When one projects that 4-year period into the later periods (say, 8- to 12-years middle management period), with normal retention there may be fewer than 50 military engineers in any 1-year group.

The SAB further pointed out that "*at the same point in time (11 years of commissioned service), the retention of military S&Es (39 percent) is even worse than that of pilots (41 percent).*"

Table 4.2: Fills of Advanced Academic Degrees versus Quotas for Air Force S&Es for FY 2000

AFIT S&E Quotas	M.S. Quota	M.S. Fills	Ph.D. Quota	Ph.D. Fills
Aero/Astro/Mech	32	10	10	0
Acq Mgt	21	20	0	0
Comp Sci/Engr	22	9	3	0
Elect Engr	43	21	8	2
Electro Optics	9	2	1	0
Eng Phys/Nuclear	17	6	8	3
Environmental	12	12	0	0
Logistics Mgmt	21	7	0	0
Meteorology	22	12	2	0
Ops Anal/Rsch	20	13	3	0
Sys Eng/Space Ops	11	2	0	0
TOTAL	230	128	35	5

Of even more serious concern to the SAB was the deficit with respect to the advanced academic degrees of S&E military officers. The

Board recommended that at least half of the S&E officers have technical master's degrees and that 15 percent have technical Ph.D.s, a level of education it deemed "*necessary to provide the technical leadership needed for high-technology acquisitions occurring within the Air Force.*" The study provided data showing advanced degree quotas versus fills for several Air Force Institute of Technology (AFIT) engineering degree programs in fiscal year 2000, Table 4.2.

As can be seen, the total fills do not come close to the quotas—128 of 230 for M.S. degrees and 5 of 35 for Ph.D. degrees. These data show

... there are serious deficiencies of the programs with respect to Air Force needs. For example, electro-optics is critical in directed-energy applications and intelligence, surveillance, and reconnaissance such as Space-Based Infrared System, the Airborne Laser, and the Space-Based Laser. Yet there are no Ph.D. fills and only two master fills. Similarly, in computer science and computer engineering, which are key to command and control and information operation application, there are no Ph.D. fills and less than half the masters needed.... Not only are officers not necessarily being educated in the right fields, but AFIT cannot be expected to operate efficiently with such low and unstable input. As a result, the Air Force will be in serious trouble technically if this situation is allowed to continue.

A follow-up inquiry by the NRC similarly drew attention to this issue, arguing that the military component of the S&E workforce should be given priority attention.[75] In addition, it found that the mix of civilian and military S&Es conveyed at least three distinct advantages:

- *Young officers entering the service bring with them fresh degrees, new perspectives, and unbridled enthusiasm that infuse 'new blood' into the enterprise. Even if these officers decide to leave the Air Force after a 3- to 4- year tour, they have served an important purpose because of the innovative element that only 'outsiders' can bring. The key is to replace these...with other newly commissioned technical officers so that the latest knowledge and the freshest thinking is a hallmark of the laboratory infrastructure.*

- *Mid-level officers who have served elsewhere in the Air Force bring a broader perspective in areas such as acquisition, logistics, and operations.... This...complements the specialized and often narrower technical perspective of the government civilian workforce.*

- *Historically, a subset of technical officers that have had laboratory experience rise to the ranks of Air Force senior leadership (e.g., general officers) and thus provide an 'S&T' perspective to the corporate decision-making process in various senior forums. Perhaps the epitome was General Lew Allen, who served in an Air Force laboratory and ultimately was selected to be the Air Force Chief of Staff. Others such as Lieutenant General Tom Ferguson and Major Generals Jasper Welch, Don Lamberson, and Fred Dropplet served in the laboratory system as junior officers, and went on to play major roles in policy, acquisition, and R&D decision-making. These officers were consciously nurtured, mentored, and promoted in an Air Force culture that recognized the value and contributions of talented young military scientists and engineers.*

"In short," wrote the NRC, *"an appropriate mix of military and government civilian S&Es, properly led and motivated, has proven its merit throughout the history of the Air Force."*

Despite the widely acknowledged value of having military officers with technical degrees in the laboratories and centers, their numbers are likely to remain relatively small until senior military leadership encourages them through promotion. At the present, having technical degrees and serving tours in the laboratories and centers are not considered "career enhancing," certainly not to the same degree as serving in an operational command. This is unfortunate because, at just the time when military-civilian partnership could bring so much to the S&T enterprise, the hard facts of career advancement discourage it.

Chapter 5

S&T: A Cost-Effective Approach to National Security

5.1 Introduction

Today, many consider the Department's S&T budget an unnecessary cost, a drain on resources better spent on things such as new weapons, logistics and supply functions, military personnel, and maintenance and repair of ships, aircraft, and facilities. Many who hold this view, however, focus only on the costs and not on the benefits, when in fact there is abundant proof that defense S&T offers a cost-effective approach to national security. Indeed, even a cursory look at this evidence reveals that even the most modest S&T investment has a very large BtC ratio.

Especially cost effective is that fraction of DOD's S&T budget expended in-house, both to carry out its portfolio of technical work and to support a small but highly trained, experienced, and readily available team of experts in a wide range of militarily important scientific and engineering disciplines. Defense dollars spent on this critical reservoir are repaid many-fold, in increased military effectiveness, avoidance of costly acquisition errors, and rapid response to threats. This in-house S&T talent is a significant source of new and future capabilities, and also provides the agility needed by today's military to act quickly and decisively. This has proved especially important in the fight against global terror, where the threat has shown itself capable of rapid evolution in response to U.S. weapons and tactics.

There are many other ways DOD investment in S&T provides significant value for money. These include innovative ideas to create budget savings that can be diverted to other high-priority areas. Often, S&T produces both near- and long-term returns on investment, not only in dollars saved, but also in lives spared. S&T produces new materials and processes that increase the reliability and safety of weapons and warfare systems such as aircraft, submarines, missiles, and torpedoes. It also creates efficiencies in operations, manufacturing processes, and maintenance—the very things critics of S&T investment want emphasized.

As discussed at length in this book, S&T investment also yields a large payback by providing the "yardstick" capability needed to avoid high cost acquisition problems. Without objective technical advice about the quality, relevance, and overall worth of work it contracts out to the private sector, the DOD cannot be a smart buyer of the products it receives. In turn, S&T spent in-house helps attract and retain the top quality S&Es who provide much of this expertise.

Regrettably, erosion of this capability has contributed to costly problems that plague an increasing number of major defense acquisition programs. Examples—in addition to the ones mentioned in Chapter Two—include the Navy's DD(X) program and the Air Force's F/A-22 Raptor. The DD(X), originally considered a $700 million cure to the rising costs of surface warships, is now projected to cost more than $3 billion each, and some estimate each Raptor could cost a quarter-billion dollars.[1] Clearly, DOD's small investment in maintaining an in-house workforce of S&Es produces positive BtC ratios when measured in such dimensions.

In other words, low investment in S&T is expensive. Ultimately, S&T innovation can not only help deter war and terrorism, but also help mitigate the high costs when they do occur. Indeed, in light of global terrorism, the cost of not investing in S&T could be terrible indeed, not only for our military forces, but for our nation.

5.2   Getting Value: It is the Positive BtC Ratios

One of the most remarkable aspects of S&T investment is its multi-dimensional payoff. S&T effort by its very nature focuses on long-term development of new technology, yet it also produces large near-term paybacks—measured in dollars and lives. In-house expertise provides a capacity to rapidly deliver solutions to battlefield problems.

The thermobaric bomb, widely credited with sparing U.S. troops the bloody prospect of tunnel-to-tunnel combat in Afghanistan, is a good example. Sustained ONR investments in high performance yet extremely survivable "internal blast" explosive formulations began in the 1980s and were brought to bear by NSWC's Indian Head Division in the mid-1990s to remedy deficiencies seen with in-service penetrator weapons used in Operation Desert Storm. Indian Head's effort produced an explosive that could survive a severe hyper-velocity penetration environment while retaining lethality in hardened deeply buried targets (in essence, the thermobaric explosives create sustained overpressures in confined spaces such as caves, tunnels and hardened structures). Because of the technical know-how resident in the DON S&T community as a result of its prior work, it could respond to a request by the USD (AT&L) and produce this capability within 60 days. How does one measure a return on investment (ROI) that saves hundreds of lives?

In fact, in the ongoing war in Iraq, there are many such quick reaction stories whose success derives from ONR's sustained investment in a reservoir of in-house S&T talent. They include improved personal armor capable of protecting the torso, face, and extremities; new detection and neutralization technology for use against IEDs; new forms of unmanned aerial vehicles to provide persistent surveillance capability; unmanned ground vehicles fitted with a variety of sensors for search and surveillance in urban terrain; and new types of vehicle armor to minimize damage radii inflicted by IEDs and other weapons. The following examples illustrate just how rapidly DON in-house capability can help U.S. forces.

- In less than four months, NSWC Indian Head produced the warhead for a new shoulder-mounted weapon needed by the Marines. The effort included a new design and fuzing system, explosive selection, initial testing, and manufacture of some 1,200 warheads for qualification and field-testing. The total effort, including deliverables, was performed within 6 months.

- In two weeks, NSWC Dahlgren produced "Dragonshield," a polymer coating that protects the Marines' High-Mobility Multipurpose Wheeled Vehicles. In early 2004, the Marines could not locate enough 3/8" protective armor, so they requested that NSWC find a way to enhance the performance of the 3/16" armor they had available. An in-house team studied performance of different polymer coatings and then performed ballistic testing to demonstrate that the new material selected did not delaminate. In fact, it enhanced the armor's performance by 50 percent. Dragonshield could be sprayed on in field with a drying time of 7 seconds, and cost only $13 per square foot. In addition, spraying a thin layer on the inside of the door reduced the thermal conductivity, preventing Marines from burning their skin on the hot armored doors.

- In response to an urgent need to supply the Marines with a heavy anti-armor capability for attack helicopters for use in Operation Iraqi Freedom, NAWC China Lake mobilized its resident know-how to develop a new metal-augmented-charge thermobaric warhead for use in the Hellfire missile. The effort, which involved total design, development, assembly, explosive loading, integration into the missile armament section, and testing, took just 13 months from funding to fielding. Secretary Rumsfeld himself commented on this new capability, noting that this new missile *"can take out the first floor of a building without damaging floors above, and is capable of reaching around corners, striking enemy forces that hide in caves or bunkers and hardened multi-room complexes."*[2]

In addition to providing rapid solutions to pressing battlefield problems, the DOD's in-house S&T investment also pays for itself many times over by improving the reliability, operational safety, and manufacturability of U.S. weapons and warfare systems. These innovations, just as those mentioned above, not only save lives, but also save money in such areas as operations and maintenance. In other words, their payoffs produce very large BtC ratios that forcefully argue the value of this small investment. Again, this point becomes clear with a few diverse examples from the DON laboratory/center community:

- A Voice Communication Processing System developed by NRL created a one-time savings of nearly $600 million to the DOD, including about $375 million to the DON. The system enhances speech intelligibility on secure telephones and provides interoperability between old and new speech parameters, allowing new and legacy phones to work together. This means 40,000 legacy units do not have to be retired prematurely, and during the potentially long transition period, secure phone service will remain uninterrupted.

- With seed money from ONR, NSWC Carderock saved the Navy $120 million in construction costs by developing and supporting certification of weldable High Strength Low Alloy (HSLA) 80 and 100 steels for ship and submarine construction. This major innovation not only reduced the acquisition cost of steel plate, but also significantly increased welding productivity by eliminating the costly process of weld preheating. In addition to using HSLA 80 in constructing the TICONDEROGA (CG 47) Class, ARLEIGH BURKE (DDG 51) Class, and NIMITZ (CVN 68) Class of ships, HSLA 100 is being used for flight deck, island, and ballistic protection structures of the UNITED STATES (CVN 75) and REAGAN (CVN 76) carriers. Savings per ship are on the order of $30 million.

- Efforts by NSWC Indian Head promise to reduce labor costs by 50 percent, increase manufacturing throughput, and improve the quality of Cartridge-Actuated Devices and Propellant Actuated Devices. These enable pilots to eject from an aircraft in an emergency. Utilizing lean manufacturing technology, the division reduced production cycle time and unit cost while improving output.

- A single innovation at NRL saved the Navy over $100 million in replacement fuel, filtering, and cleanup costs, and increased operational safety and combat readiness of Navy vessels. In-house S&T personnel invented a method for assessing distillate fuel stability, an innovation that has reduced the number of incidents in which vessels have shut down or failed to achieve full power because of contaminants, which result from chemical reactions in fuels stored for extended periods of time. This method has also been adopted as an American Society for the Testing of Materials Standard.

Although it is sometimes possible to estimate the dollar savings an innovation yields, as these examples illustrate, more often this is not the case. In fact, there has been no systematic effort to quantify such benefits because there is no methodology for generating comprehensive BtC ratios.[3] It should be emphasized that BtC is more comprehensive than traditional ROI approaches, typically employed by industry and measured only in dollar savings. As demonstrated above, defense S&T innovations often produce an array of benefits, many of which are not captured in a simple financial ROI.

To help remedy this deficiency, ONR's N-STAR office commissioned a study[4] to develop a more comprehensive BtC methodology. Using the Navy's ILIR (discretionary 6.1) program as a test case, the idea is to use the results to ascertain the total benefits derived compared to the total funds invested. NUWC Newport Division was selected for the pilot study. The division has operated a modestly sized ILIR program for more than 30 years, and has a detailed database on projects funded along with their history of transition into follow-on torpedo development efforts. The study, overseen by the ILIR program manager along with NUWC S&Es and other Navy principals, has already examined several projects. One involves using ILIR-generated technology to increase the MK-14 ADCAP (Mod 5) torpedo's stealth. The knowledge base used in this upgrade is largely the result of a 20-year investment in some 33 ILIR projects totaling $4.7 million (in inflation adjusted FY 2005 dollars), and has produced a substantial corporate memory at NUWC that helps the center transition S&T into development programs.

The results of this case study impressed even those who fully expected impressive results. The Mod 6 upgrade version reduced radiated noise by more than two orders of magnitude. Importantly, it was ILIR-funded research that enabled this reduction without a replacement of both the energy and propulsion sections of the torpedo. Of the total estimated savings to the Navy of $1,558 billion, about 80 percent or *$1.247 billion* is attributable to the ILIR effort alone. Comparing this to the $4.7 million investment produces a BtC ratio of 264:1. Even comparing the $1.247 billion to the entire $114 million (in inflation adjusted FY 2005 dollars) ILIR investment (not just the portion focused on torpedo stealth) at NUWC from 1971 to 2005 produces a BtC ratio of 10:1. ILIR technology in this effort also produced several other benefits, such as accelerated Fleet deployment of the Mod 6 torpedo. Over 70 percent of the total ADCAP inventory has been converted to Mod 6 to date.

Three additional ILIR efforts at NUWC are now being studied, including submarine towed arrays and a fuel cell project aimed at future unmanned undersea vehicles. A comprehensive draft report on this new BtC methodology has been delivered to the N-STAR office for review. If this pilot effort continues to show promise as a means of accurately and comprehensively estimating the overall BtC of ILIR-funded programs, it will be extended to other DON warfare/system centers that receive such funds.

5.3   The High Cost of Not Acting

Just as there are substantial benefits derived from DOD's S&T investment, there are costs incurred as the result of investing too little. It is useful to consider just how high these costs can be, measured in lives lost, injuries incurred, and depletion of our national resources.

To help put the DON's annual S&T investment (currently at some $1.8 billion) in some perspective, it is useful to examine the enormous costs of military conflicts and terrorist acts. Some of these costs are direct, for example the impact on the budgets of the countries involved, while others are indirect, for example the overall effect on the world economy. Of course there is a multitude of situational factors. To give just one example, costs generally depend on the size of the conflict and its duration, although a short conflict can certainly involve very high costs, as the case of a nuclear exchange makes obvious. Because of the compounding effects produced by an almost countless number of situational factors, estimating the costs associated with war is replete with difficulty and involves tremendous uncertainty. Even so, some analysts, such as William Nordhaus of Yale University, have tried. In one such example, he analyzed the likely costs of the Iraq war. As part of this effort, he also estimated the direct military costs of past American wars, from the Revolution up to the first Gulf War, Table 5.1.[5] While these estimates do not include veterans' benefits and health costs, which

Conflict	Total Direct Costs of Wars (Billions of Dollars)		Per Capita Cost (Constant 2002 Dollars)	Cost (% of GDP)
	Current Year Dollars	Constant 2002 Dollars		
Revolutionary War (1775-1783)	0.1	2.2	447	63
War of 1812 (1812-1815)	0.09	1.1	120	13
Mexican War (1846-1848)	0.07	1.6	68	3
Civil War (1861-1865)				
Union	3.2	38.1	1,357	84
Confederate	2.0	23.8	2,749	169
Combined	5.2	62.3	1,686	104
Spanish-American War (1898)	0.4	9.6	110	3
World War I (1917-1918)	16.8	190.6	2,489	24
World War II (1941-1945)	285.4	2,896.3	20,388	130
Korean War (1950-1953)	54.0	335.9	2,266	15
Vietnam War (1964-1972)	111.0	494.3	2,204	12
First Persian Gulf War (1990-1991)	61.0	76.1	306	1

**Table 5.1: Costs of Major American Wars
(Source: Nordhaus, Table 2)**

are very large indeed for more recent wars (a point developed below), they certainly indicate that major wars have cost a lot in comparison to GDP estimates at the time of the conflict. The costs of any war are high, and the costs of losing are even higher. John Holzrichter, an assistant to the director of the DOE's Lawrence Livermore National Laboratory, states, "In my opinion, nothing can be worse than losing a conflict of the magnitude of a world war. Even the so-called winners of past conflicts faced immemerable negative consequences, especially our European allies."[6]

Obviously then, avoiding a war offers enormous savings. Holzrichter continues, "Of course, in more recent history, a nuclear exchange would have been a worldwide catastrophe. I firmly believe that the prevention of such conflicts over the past 55 years, using military deterrence, information dominance, and diplomacy, has been the greatest success in America's national security history."[7] The deterrence Holzrichter mentions was made possible by maintaining a technological superiority over would-be adversaries. Today, as throughout the Cold war, U.S. technological superiority depends on a vibrant national S&T effort, including a healthy and productive in-house S&T component. Calculating cost of the Global War on Terror is even more difficult. The elements that contribute to the overall costs of terrorism are more varied than with a traditional war. In addition to the costs from the immediate aftermath of the terrorist act, there is a continuing economic impact, such as the added cost to protect citizens from further attacks. This "terrorism tax" is high, and imposed on every citizen.

Regardless of the difficulties, many are attempting to determine the costs of terrorism. Consider Table 5.2, taken from a study by the Milken Institute.[8] Simply the sheer number of cost elements shows that estimations are fraught with complexity and uncertainty. Many have attempted to assess the overall costs associated with 9/11. By one estimate, the real and human capital costs ranged from $25 billion to $60 billion.[9] Another put the human capital loss at $40 billion, and the property loss between $10 and $13 billion.[10] Still another claimed a total direct loss of about $21.4 billion.[11] While such estimates are imprecise, they certainly hammer home the point that even a small shadowy network of individuals can inflict tremendous costs on a nation, and for very little money.

As mentioned, another high price of war is the veterans' benefits and associated health costs, which are particularly large for recent conflicts. One reason is that, even though the firepower today has increased, the lethality of wounds has decreased, as the result of improvements in the military medical system. This has especially been

**Table 5.2: Cost Categories for Terrorist Attacks
(Source: Navarro and Spencer, Exhibit 1)**

The Immediate Aftermath	Longer Term Microeconomic Effects of a "Terrorism Tax"	Sector-Specific Impacts	Government Bailouts and Budgetary Impacts	The Higher Oil Prices-Weaker Dollar Conundrum	Macroeconomic Costs
- Property damage - Loss of human life and injuries - Loss of economic output - Reduction in stock market wealth - Psychological impacts of terrorism	- Increased airline security - Other security measures	- Advertising - Airlines - Insurance - Hotel and tourism	- The airlines - E.g. New York City - Reduced federal, state, and local tax revenues	- Effects of an oil price shock - Costs of a weaker dollar - Destabilization of the stock market and international monetary system	- Recession - A more volatile business cycle - A lower long-term growth pattern and economic stagnation

the case since the end of the Persian Gulf War, and includes changes in strategies and systems of battle care. As reported by Gawade, "In World War II, 30 percent of the Americans injured in combat died. In Vietnam, the proportion dropped to 24 percent. In the war in Iraq and Afghanistan, about 10 percent of those injured have died. At least as many U.S. soldiers have been injured in combat in this war as in the Revolutionary War, the War of 1812, or the first five years of the Vietnam conflict, from 1961 to 1965."[12] These numbers and others in Table 5.3 illustrate how the lethality of war wounds has changed over time.

Table 5.3: Lethality of War Wounds among U.S. Solders (Source: Gawande based on DOD data)

War	No. Wounded or Killed in Action	No. Killed in Action	Lethality of War Wounds
Revolutionary War	10,623	4,435	42
War of 1812	6,765	2,260	33
Mexican War	5,885	1,733	29
Civil War (Union Forces)	422,295	140,414	33
Spanish-American War	2,047	385	19
World War I	257,404	53,402	21
World War II	963,403	291,557	30
Korean War	137,025	33,741	25
Vietnam War	200,727	47,424	24
Persian Gulf War	614	147	24
War in Iraq and Afghanistan, 2001-Dec 2004	10,369	1,004	10

Aside from the personal tragedy involved here, these numbers also translate into very large health costs. The number of killed or wounded in action shown in this table has of course increased, and as of 4 November 2005 stands at 17,067, an increase of almost 65 percent.[13] Of the wounded, just under half return to action within 72 hours. But over half do not return to duty quickly, and many are permanently disfigured and disabled. One recent story reported that the U.S. Senate had approved $1.5 billion in emergency funds for veterans' health care to pay for the rising cost of Iraq war injuries.[14] In June 2005 testimony before the House Veterans Affairs Committee, James Nicholson, Secretary of the Department of Veterans Affairs, noted that 2002 projections estimated that 23,553 veterans from the Iraq and Afghanistan wars would need health care, but this figure has been revised to 103,000. As a result, Nicholson said, health care costs were rising at a pace of 5.2 percent, far beyond the 2.3 percent annual growth originally projected.[15]

Whatever the exact numbers, the costs of war-related wounds are very large and at the moment growing rapidly. The IEDs causing so much damage to U.S. troops in Iraq have created an urgent demand for new technological breakthroughs to counter them or mitigate their impact. Such breakthroughs are precisely the role of in-house S&T efforts carried out by S&Es working in the DON laboratories and centers. The BtC ratio of finding a technological solution to the IED threat alone would be enormous. The point here is not to determine the costs associated with the war or terrorism, but to further underline the great value of maintaining a strong defense S&T effort, and especially to affirm the necessity of revitalizing a DON in-house S&T enterprise teetering on the brink of becoming inadequate for the tasks placed before it. In terms of the overall defense budget, the investment in S&T is relatively modest, but one that produces large payoffs.

Chapter 6

A Road Map to Action

6.1   Introduction

The U.S. is today the sole world superpower. It enjoys a position of unparalleled military strength and uses that strength to defend itself and its allies. Indeed, providing for the national defense is the first and fundamental commitment of the Federal Government. But the task of national defense is rapidly changing as it must in order to counter evolving threats such as global terrorism. As it has since the end of World War II, the country must equip the men and women who comprise its armed forces with superior technology that enables the best possible warfighting capability. Leveraging the products of S&T to provide this technological edge is still a valid strategy for a new century. It is in fact an absolutely essential strategy, as our forces face an ever-growing array of threats from a determined enemy. The U.S. must sustain a vigorous national defense S&T effort not only in the private sector, but in the DOD's own in-house laboratories and centers as well.

Despite the growing indications that the DOD will need it, this in-house S&T workforce has been allowed to atrophy both in size and technical quality. And the vigorous in-house S&T stabilization and rebuilding now required faces significant challenges. Local, national, and global trends all indicate that staffing with the right kinds of S&Es will not be easy, especially in light of the rapid build-up of scientific talent and infrastructure outside the U.S. Many of the trends previously discussed have already constricted the flow of talent in the pipeline that feeds DOD's hiring pool. Further, global trends have changed the very manner in which innovation occurs, all of which means the DOD must reshape its S&T enterprise into one suited to 21st century needs. Realizing this new vision requires achieving four overarching goals:

- First, there must be a sufficiently large community of S&T "prospectors" who can participate as peers with colleagues throughout the global community, thereby ensuring that the fruits of the worldwide enterprise can also be applied.

- Second, the DON must develop and implement a human capital strategy for its in-house S&T community that will work on today's most important problems. The strategy should result in 4,000 new Ph.D.s by 2015.

- Third, S&T funding should be increased to 3 percent of overall DON total obligation authority, and it should be maintained at that level in constant, inflation-adjusted dollars through 2015 to ensure revitalization of the workforce and provide the intellectual capital base for the Navy-After-Next.

- Fourth, the DON must support and emphasize technical excellence, and appoint visionary civilian leaders who have authority and responsibility for the technical output of their organizations.

To realize these overarching goals, the following sections will provide 10 recommendations for revitalizing the defense S&T enterprise. Because the authors are especially familiar with the DON and have access to much of the relevant data, a considerable portion of what follows will use that Service as a framework. However, given the great similarities among the DON, Army, and Air Force in-house S&T management issues, many of the steps recommended here could apply to those Services as well.

6.2 The Global Connectivity Imperative

Recommendation 1: Expand ONR Global activities to include all the major players on the international S&T scene.

As S&T breakthroughs become more and more the product of collaborators often scattered around the world, the DOD is becoming a relatively small player in the global enterprise. To track, assess, and apply the products to address new warfighting needs, the DOD will have to open its window wide. Put another way, it must engage the enterprise on a global scale, staying connected to all relevant prospecting communities regardless of location. To do this, members of the workforce must be "card-carrying" members of those communities—credible publications are the ante for playing at the global S&T table.

Recognizing this need, ONR has reached out to the world S&T community through its ONR Global organization, with offices in London, Santiago, Tokyo, Australia, and Singapore. They employ some 40 scientists, technologists, and engineers to help work with R&D communities in a broad range of technical areas worldwide to provide solutions to Naval challenges. The S&T division engages academia, defense and commercial industries, and government agencies. It encourages information exchange and collaboration with international scientists and organizations whose interests are of mutual benefit. At present, the division utilizes several vehicles, including the Visiting Scientist Program (VSIP), the Conference Support Program (CSP), and the Naval International Cooperative Opportunities in S&T Program (NICOP).

- The VSIP supports short visits among DON personnel, contractors and grantees, and international S&Es to explore collaborative opportunities, exchange information between DON and international programs, assist ONR Global with international liaison and assessment, and establish long-term relationships between DON and international funding agencies or next-generation international S&T leaders.

- The CSP financially supports workshops and conferences to plan collaborative international programs, exchange information between DON and international programs, identify and discuss issues of interest, develop relationships between DON and international S&T sponsors, and maintain international field office connectivity in core areas.

- The NICOP encourages international collaboration by providing seed funds when ONR headquarters or other U.S. government agencies commit to outyear program funding. Priorities include supporting transformational initiatives and accelerating the introduction of breakthroughs into naval applications.

These efforts are a good start, but given the rapid pace of technological globalization, bolder steps are needed. ONR Global should be expanded dramatically to include sites in other countries emerging as major players, for example India and China. New forms of engagement could also be utilized. For example, scientists in former Soviet Union countries could be used to screen and assess technology flowing out of Russia, China, and India. For a relatively modest investment, highly skilled and highly educated S&Es in these countries could be employed to do work themselves and to provide a first line of defense against technological surprise.

To reinforce and institutionalize this imperative, the DON should formalize the idea that S&Es must be globally connected. Inserting appropriate language in the S&T strategy would help accomplish this, and also help ensure resources are available for the workforce to participate in global prospecting communities. Additionally, individual development plans could include requirements to further global connectivity. In today's business-oriented environment, foreign travel and attendance at international conferences are many times discouraged, but this climate must change for members of the S&T community to engage their international technical communities.

6.3   A New Human Capital Strategy

Recommendation 2: The DON should bring on board 500 S&Es per year over the next ten years to pursue research and technology areas of critical importance to developing future military capabilities.

Data previously presented demonstrated the shrinking of the DON in-house S&E workforce since the end of the Cold War, including that subset focused on S&T. In addition, that workforce has experienced a significant brain drain due to the loss of many of its most highly educated and knowledgeable members, a fact reflected in the drop in the number of Ph.D.s. To meet the increasing demands likely to be placed on this workforce, the DON should immediately act to stabilize its size and strengthen its quality.

Exactly how many S&Es the DON needs is of course dependent on many factors, some of which are either unknown or unknowable. They include things like forecasts of future workloads, attrition rates, and requirements in terms of education, training, and skills. The lack of solid information about these and other factors hinder accurate projections.

Nevertheless, there is enough information to produce usable estimates, and such estimates were the goal of a recent study the N-STAR office commissioned to help with workforce planning.[1] Utilizing an Office of Personnel Management workforce planning model, the study compared the current situation with estimated future requirements to identify hiring needs. Estimates of attrition rates were projected based on historical loss rates (about 20 percent per year). However, projecting future demands is more difficult owing in part to inconsistent and unpredictable congressional funding patterns. The N-STAR study effort was also limited in that its authors could obtain funding and attrition data only from a subset of the overall DON laboratory/center community's S&T enterprise (the NAWC, NUWC, and the SSC). To help overcome these uncertainties, the approach utilized best- and worst-case scenarios and other scaling methodologies.

Despite such difficulties, the ONR study authors combined their original research with other information to reach several conclusions. One, of the NLCCG community's current population of some 21,000 S&Es, about 4,000 of them work on S&T programs, including about 1,900 who have Ph.D. degrees.[2] Of these, a smaller subset consists of a highly talented group of so-called "Esteemed Fellows"[3] that constitute an in-house cadre of intellectual capital responsible for a large portion of the future innovation the DON will need to support naval operations. These are much of the brain trust of the laboratory/center community, and its leading prospectors.

In addition to ensuring an adequate size of the S&T workforce, another aspect of revitalization that needs attention is refreshing it periodically to secure the right mix of technical capabilities. Staying abreast of today's technology base requires constant acquisition of new skills and knowledge. This in turn requires a steady workforce shaping, which can be achieved in part through planned personnel turnover. In other words, a healthy, cutting-edge S&T workforce should exhibit a healthy rate of turnover. It is reasonable to expect that the entire workforce should turn over about every 10 years. This estimate is linked to academic refresh cycles, the currency of technology developments, a flexibility in the retirement system that encourages movement into and out of the federal sector, and the fact that technologists tend to migrate with projects into follow-on programs. In any case, turnover creates vacancies that must be filled on an on-going basis.

Given the estimated need in the NLCCG community for some 4,000 S&Es, a 10-year refreshment cycle, and projected rates of turnover of about 100 each year, it is reasonable to assume that about 500 will have to be hired each year to build and maintain an in-house workforce.

The increased flow in the pipeline required for this many new hires necessitates a comprehensive human capital strategy, and unfortunately, DOD's current system is unsuitable for today's high technology organizations. Several recent studies have in fact commented on this, and recommended approaches equal to the task. A basic problem, as discussed in Chapter 4, is that one-size-fits-all approaches like those embodied in the CSS and the NSPS are simply too rigid for the dynamic needs of the laboratories and centers. In fact, many within the defense technical community fear the NSPS may actually take away many of the flexibilities and authorities the laboratories currently enjoy under various congressionally established personnel demonstration programs. The

DDR&E, other senior leaders of the OSD, and Service technical communities must ensure that the final version of the NSPS offers the centers not only what they have today, but in fact the additional flexibilities to operate as 21st-century technology enterprises.

A comprehensive human capital strategy for the laboratories and centers must incorporate a number of features. It must facilitate the hiring, training, and retention of the best technical talent, especially those individuals in cutting-edge efforts. It must foster acquisition of advanced technical degrees, interdisciplinary education, and continuing education. It must also focus on knowledge management (KM) issues, especially how to transfer a massive corporate memory from Baby Boomers to the next generation.

All this will of course require sufficient resources. In fact, the N-STAR initiative has several components aimed at increasing the flow of new talent into the hiring pipeline, a full description of which is given in Appendix A-1, but money will also be required to fund these new employees. In fact, providing an assured level of funding for their first two years of employment is an incentive to accept a position and a magnet to help retain them.

In addition to the activities noted under the N-STAR umbrella, several other DOD-wide initiatives are geared to help fill the S&T recruiting pipeline. Prominent among these is the SMART scholarship program mentioned in the Introduction. Established by Congress in the FY 2005 National Defense Authorization Act, SMART promotes the education, recruitment, and retention of upperclassmen and graduate students in science, mathematics, and engineering. All scholarships are based on individual DOD laboratory/center needs. In FY 2005, Congress provided more than $2 million for scholarships of up to two years.

Impressed with the potential of the SMART initiative, Congress is expected to follow it up with the National Defense Education Program (NDEP) for 2006. Already, during consideration of the FY 2006 defense appropriations bill, the Senate agreed unanimously to an amendment to provide an additional $10 million for NDEP. It is anticipated this program will provide scholarships toward attaining a single degree, from the Associate's to a Ph.D.

Figure 6.1 suggests a balanced approach for hiring the 500 new S&T workers needed each year. As discussed, the new personnel should either possess a Ph.D. or be on an educational track geared toward that level. The underlying concept is as follows. First, there would be 250 S&Es coming out of two-year undergraduate scholarship programs (DON funding for junior and senior years with payback requirement). Of these, 100 will feed directly into the S&T workforce with the other 150 allowed to move directly into one-year M.S. programs. At that level there will be funding for 250 students per year. Of these, 150 will come from the B.S. pool just described, while the other 100 will come from DON warfare/system center employees and former military officers who want to pursue graduate education. Of these 250 M.S. graduates, 200 will move into the centers each year to fulfill their service obligations. The top 50 can proceed directly to Ph.D. programs. At that level there will be 200 new employees each year. The pool will include the 50 M.S. graduates with the balance of 150 coming from a combination of post-doctoral and other stipend programs.

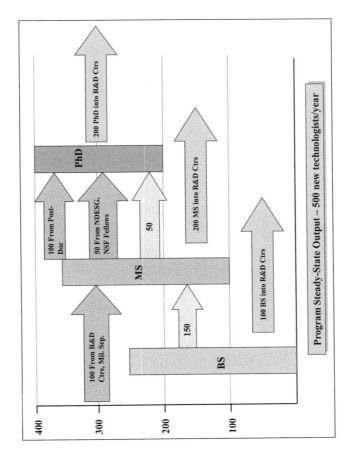

Figure 6.1: Navy S&T Hiring Pipeline

Recommendation 3: Institute a DON S&T community expectation that career path to journeyman level requires a Ph.D. or equivalent education.

As discussed in detail in Chapter 4, the nature and rate of technological change today means those planning S&T careers will need a more highly, broadly-based, and multidisciplinary education than ever before. This situation has already prompted the National Academy of Engineering and other groups to recommend educational changes for the new generation of S&Es. In fact, engineering organizations have recommended significant changes such as more interdisciplinary learning, and the NAE and others have suggested that a B.S. signifies an "engineer in training," while an M.S. should be considered the "professional degree."[4] Indeed, because *"the half-life of cutting-edge technical knowledge today is of the order of a few years,"* universities need to offer advanced training to working engineers seeking to maintain their technical relevance.[5]

These requirements are just as applicable to the defense laboratories. Indeed, if the M.S. is the ante for getting into the DON's S&T workforce, then a Ph.D. or equivalent should be the goal for each professional, especially those who aspire to become Esteemed Fellows. For that reason, new hires at the centers that do not already possess a Ph.D. should be expected to enter a career path to achieve that degree or its equivalent to reach the professional level.

While the goal should be a S&T workforce comprised wholly of Ph.D.s, the reality is that many new hires will enter the workforce below that educational level. Therefore, each of the laboratories and centers must offer a vigorous program of continuing education opportunities. At a minimum, all S&T workers should receive the equivalent of a Ph.D. through various educational and on the job training opportunities.

Examples of how this continuing education could be provided have been given. They include distance learning methods as well as collaborative ventures with colleges and universities, such as the one between NSWC's Indian Head Division and the University of Maryland's CECD. In addition, new hires should also be expected to remain connected with academia, either as a student, teacher, or participant in some joint outreach effort aimed at creating and sustaining more interest in science and engineering as an occupation. N-STAR's "Virginia Demonstration Program" is an excellent model for such outreach. S&E staff at DON warfare center sites in Virginia mentor students and collaborate with teachers. The program has the principal goal of generating enthusiasm among 7th graders in science, engineering, and mathematics. It also aims to invigorate the science and math curricula in Virginia schools and contribute to the professional development of middle-school science and mathematics teachers, while developing a number of Ph.D. researchers in fields of interest to the DON. Such activities should be part and parcel of individual development plans for S&T community members.

Because women and many minorities are under-represented today in many fields of science, engineering, and mathematics, special care must be taken to increase their participation. It is simply in the national interest to do so. Several efforts with this goal are already underway, for example the DOD's Historically Black Colleges and Universities/Minority Institutions program. Also, the public-private partnership BEST (Building Engineering and Science Talent) was launched in 2001 at the recommendation of the Congressional Commission on the Advancement of Women and Minorities in Science, Engineering and Technology Development. BEST convened some of the nation's most respected practitioners, researchers, and policy makers to identify best practices for developing the technical talent of under-represented groups, from pre-K through higher education and the workplace. The results can be found in the several reports linked on the BEST web site. [6] Other initiatives should also be undertaken, and long-term relationships between minority academic institutions and DON technical organizations cultivated. Importantly, all outreach programs should be fully integrated to attain their maximum benefit.

Recommendation 4: Create a DOD S&T Academy, equivalent in prestige to National Academies in Science and Engineering.

As discussed in Chapter 4, because there are limited opportunities for membership in the national academies for those in the DOD's in-house S&T workforce, the Department needs some institutional mechanism to recognize its top researchers for their outstanding contributions to national defense. One solution would be for the Department to establish an S&T academy that carries an analogous recognition. Election to membership would entail public recognition by top national and DOD leaders and various prizes or other privileges. For instance, election could provide a stipend for work on other pressing technical problems, or a sabbatical for work with outstanding researchers. This would of course require an appropriate funding line. In fact, one could envision members of this select fraternity being allocated funds to develop teams of researchers working on particularly difficult defense problems, with the long-term result of developing a pool of new, prospective members.

A good illustration of the necessity of appropriately resourcing such a program is the authority granted to the three Services in the National Defense Authorization Act for 2001. The provision, Section 1113 (*Extension, Expansion, and Revision of Authority For Experimental Personnel Program for Scientific and Technical Personnel*), extended to the Services the same authority previously granted to the Defense Advanced Research Projects Agency. In effect, the authority allowed the agency to carry out an experimental five-year program to *"facilitate recruitment of eminent experts in science and engineering for research and development projects."* More specifically, it allowed appointment of

20 S&Es from outside the civil service and uniformed Services "*without regard to any provision of title 5, United States Code, governing the appointment of employees in the civil service.*" Essentially, Section 1113 gave each of the Services 40 such positions. However, Congress provided no funding to support the incumbents. As a result, there have been few appointments, rendering the authority of little use.

6.4   Resourcing the DON S&T Enterprise

Recommendation 5: Increase the annual DON S&T budget to 3 percent of TOA, and reallocate amongst community members (academia, in-house centers and industry) to ensure viability of each sector.

The new S&T enterprise must have adequate financial and human resources. The financial resources must support the in-house technical work and initiatives aimed at attracting, training, and retaining new employees. Today, the DON S&T workforce includes approximately 4,000 S&Es.[7] To revitalize this workforce and build the skills needed for the future, the DON should hire about 500 new S&Es each year, emphasizing Ph.D.-level talent. As noted in Chapter 2, discretionary dollars in the hands of TDs greatly enhance the recruitment and retention of high-quality, productive personnel. Consequently, a suitable portion of funding should be distributed at the discretion of the TD for work deemed important. Again, the resources should come from stable funding sources, because most S&T efforts span years. Funding instabilities disrupt program planning and execution, compel the workforce to fritter away precious time chasing new funding, and lead to erratic hiring patterns that disrupt strategic workforce planning.

Despite overwhelming evidence that relying on the private sector is risky business, many policy makers still prefer to believe it can be done, and why not? After all, almost daily the news seems to announce yet another innovation. This implies that most innovations come from commercially funded R&D rather than some defense laboratory. In fact, Americans enjoy a flood of new products made possible by innovations in the electronics, materials, information, and health fields. Many of these have also transformed military operations—advances in communications and electronics, for example, have enabled network-centric warfare. Not surprisingly, the commercial sector's performance has impressed military planners, especially those involved with front-line military operations who are, after all, only interested in functional capability. Besides, it is easier to see a new piece of hardware than a new idea about nanotubes.

Although the commercial sector pumps a great deal of money into innovation, is it planting the seeds that grow into this seemingly constant harvest of new technologies? An illuminating and forcefully argued recent paper called *The S&T Innovation Conundrum* helps answer that question.[8] Importantly, it also demonstrates the risk inherent in a defense strategy that relies too heavily on the private sector for breakthrough technologies.

Surprisingly, the authors found that a breakthrough innovation takes 15-20 years to progress through the early, or "prospecting" phase—the same amount of time it took 100 years ago. In examining the histories of some major S&T innovations over the last century, they found that the nature of human creative timescales is rate limited. In other words, technology has not sped up the rate at which scientists process information and create new knowledge. The only way to create new knowledge faster, then, is to fund more prospectors. This will enhance the probability, but not the certainty, of discoveries.

With a compelling distinction between this early prospecting phase and the later "mining" phase, the authors show how technological advances are largely an extraction (mining), the success and speed of which depend on the long-term heavy lifting of the prospecting. Indeed, there is often no useful functional capability produced during the prospecting phase, which is why few companies risk their capital on basic and applied research that may not provide a return on their investment. Instead, most capability results from well-funded commercial sector development programs that exploit the prospecting. Further, the mining phase is not rate limited, but can be sped up.

Lack of awareness about the differences between these phases has fueled misperceptions about funding defense S&T. The authors note that many DOD planners incorrectly assume that rapid advances in a general functional capability also represent the tempo and time scales for breakthroughs in the prospecting phase. Additionally, few are aware that if more prospecting is not done now, the miners in 20 years will find a barren science landscape from which to draw.

It is important to note that while both the prospecting and mining phases require basic and applied research as well as exploratory development, they involve different mixes. Therefore, efficiency requires different forms of governance. That is, each phase requires its own approach for allocating resources such as people and dollars. As the authors explain:

Economists, understanding the important role that technology plays in economic growth, have begun developing theories regarding the impact on economic growth where R&D investments are determined so as to maximize profits. This is referred to as an endogenous investment strategy. These theories help in discussing how economic conditions combined with an endogenous investment strategy for R&D affect the scientific and technical talent pool and the generation of knowledge as well as economic growth.... It seems clear that a solely endogenous approach to determine R&D investments results in too little long-term research being funded. Talent and resources gravitate to the mining phase at the expense of the prospecting phase and at the expense of knowledge generation needed to sustain economic growth in the long term.

Simply stated, if economic forces alone dictate allocation of resources, then those resources will migrate to where money is to be made from innovation that produces new functional capability, because that is what grabs market share. No wonder commercial firms today (including defense companies) shun long-term, high-risk prospecting ventures—they simply prefer that someone else underwrite them. In fact, only the Federal Government has both the motivation and the deep pockets to fund most basic and applied research. This is all the more the case with S&T spending for national defense.

The next question, then, is what form of governance will achieve a satisfactory return on investment, especially considering that in this early prospecting phase it is impossible to predict exactly where innovations will emerge. Increasing the number of prospectors certainly increases the probability of breakthroughs, but the undertaking still involves a large dose of serendipity. So what is the best course? Here the authors provide excellent advice:

Organizations that are in the business of actually making things or actually using things need to staff their organizations with this in mind. Since information technology has not reached the point where it can or should replace human creativity it is necessary to actually engage the relevant prospecting communities. The most effective way to do this is to have staff who are card-carrying members of those communities. This is necessary in order to understand what the state of the art is and to have access to it. If one is not a card-carrying member one gets no respect in such communities, and therefore no serious entry. Furthermore, since science and technology is a global undertaking (and becoming more so) it is essential to engage on a global scale. This is true for DOD as well as commercial businesses.

The U.S. is becoming a relatively smaller player in this global enterprise. It is not possible for any one organization, including DOD, to conduct science and technology effort in all areas of this global science and technology undertaking. It is possible, however, for large organizations like DOD to employ a large enough science and engineering workforce to gain entry into most areas of prospecting space. This will provide the best window on what is going on out there and for evaluating its importance. The strategy should be to have a workforce that has the stature to be welcome and the broad competence (scientific and military) to recognize something that is important to DOD when it sees it. This should be DOD's science and engineering brain trust.

Such a "brain trust" is critical to future defense needs, and must be revitalized now and treated as the precious commodity it is.

In the DON, the in-house S&T workforce constitutes this brain trust—just how many workers are needed for it to be adequate? For the reasons just mentioned, it cannot be large enough to do everything on its own. Still, it must sustain a presence in all of the prospecting communities important to the DON. Only those members who do hands-on S&T can acquire the bona fides to gain entry as prospectors; contract monitoring of outsourced work will not suffice.

Given, then, that the primary job of the DON in-house S&T workforce should be prospecting, not mining—a task better left to the commercial sector—just how many workers are needed? As stated, the current best estimate is around 4,000. Moreover, accounting for annual workforce attrition rates, sustaining and refreshing this number probably require about 500 new hires per year over a 10-year cycle.

Providing adequate resources for the in-house community and enhancing the prospecting component of the S&T portfolio—without jeopardizing industry's participation—requires a budget increase. The job simply cannot be done with existing resources. The 2001 Quadrennial Defense Review argued, *"DOD should maintain a strong [S&T] program in order to support evolving military needs and to ensure technological superiority over potential adversaries."* Doing this entails an *"increase in funding for S&T programs to a level of three percent of DOD spending per year."*[9] Recall that several DSB studies also set this three percent goal.[10]

The Congress too supports a robust defense S&T program funded at this three percent goal:

- *"Our present military strength is the result of substantial S&T investments made a generation ago…. In a similar vein, our nation's prospective security and military dominance ultimately depends on its ability to perpetuate technological advantages over the next few decades. S&T programs will enable us to maintain this advantage…. It is imperative, therefore, that we act to fund S&T at 3 percent of the total defense budget."*[11]

- *"The committee commends the Department of Defense commitment to a goal of 3 percent of the budget request for the defense science and technology program and progress toward this goal. The committee views [such] investments as critical to maintaining U.S. military technological superiority in the face of growing and changing threats…and believes that both the defense agencies and the military departments have vital roles in DOD's science and technology investment strategy."*[12]

- *"The Committee feels that a robust defense science and technology program is a requirement in order to develop the new systems and operational concepts that will enable transformation…. The Committee fully supports the Department's stated goal of investing 3 percent of the defense budget in science and technology programs [and]* **urges the Department and each of the military services to achieve the 3 percent goal as soon as practicable.**"[13] [Emphasis added]

Given the strong, consistent, widespread support for this goal, funding annually at 3 percent of DON S&T is recommended. The DOD and the Services should strive to achieve it as soon as practicable, particularly because, as Chief of Naval Research Cohen himself recently stated, *"In the war against terrorism, S&T is the enabler which links innovative research to warfighter and homeland defense requirements."*[14] Under-funding such a vital requirement is not tenable at this critical juncture.

Given that laboratories and centers conduct research in all three phases of S&T (BA 1, BA 2, BA 3) and that they are part of a three-partner team that includes academia and industry, how should the roles be balanced among the three categories? Chapter 2 discussed the partners in the defense technology base, noting that DOD has historically relied on universities and non-profits for the largest share of its basic and applied research. The arrangement allows access to some of the world's best minds and newest ideas, and also to facilities DOD does not have to support unilaterally. It also engages students in defense work, thereby providing access to the next generation of technologists, engineers, and managers. Industry, on the other hand, is well suited to work in the later stages of the RDT&E process. Recently, it has had a growing role in S&T as well, for example by participating in large demonstration and evaluation efforts.

The roles of the DOD in-house laboratories and centers are based on their fundamental duty to support a broad range of national security efforts. For example, they are best able to translate between technological opportunities and the warfighter's needs, integrate technologies across life cycles and generations of equipment, respond rapidly to warfighting needs, provide facilities that private companies cannot or will not, and offer the Services the yardstick capability necessary to become smart buyers and users of technology. To satisfy these varied needs, they must carry out S&T effort in all its phases as well as participate in a full spectrum of RDT&E activities.

S&T Budget Activity	Sector	Current Distribution	Recommended Distribution
BA1	Universities and Non-Profits	54	55
	In-house Laboratories/Centers	33	36
	Industry	13	9
BA2	Universities and Non-Profits	21	22
	In-house Laboratories/Centers	35	56
	Industry	44	22
BA3	Universities and Non-Profits	19	13
	In-house Laboratories/Centers	20	25
	Industry	61	62

Table 6.1: DON S&T Performance by Defense Technology Base Partner

Table 6.1 shows the current and recommended distribution of S&T among the private and public sector partners. Based on FY 2004 data, universities and non-profits perform the largest share of basic research (BA1) (54 percent), while industry predominates in applied research (BA2) (44 percent) and advanced technology development (BA3) (61 percent). Of total actual dollars received by the different participants aggregated across all three budget categories, universities and non-profits received $614 million, in-house laboratories/centers $624 million, and industry $1,011 million.

The table 6.1 suggests what the authors believe would be a more appropriate distribution. The current situation regarding applied research undercuts the ability of the in-house laboratories to bring new ideas to fruition in new military capability. Centers could be more productive at this transitioning if they received an increased share of applied research funding. As recommended, universities and non-profits would still predominate in basic research (55 percent) and industry in advanced technology development (62 percent). However, the share of applied research (BA 2) in-house would increase from 35 to 56 percent. Using a FY 2004 TOA of $115 billion, the three percent recommendation and this suggested distribution would result in the following funding totals: Universities and non-profits ($993 million), in-house laboratories/centers ($1,335 million), and industry ($1,123 million).

The net result of this change would be a significant increase in the budget aimed at the prospecting phase of innovation the commercial sector currently ignores, accompanied by a strengthening and revitalization of the overall in-house S&T enterprise. Finally, it would institutionalize the idea that each sector of the S&T community has a "power alley". For academia, non-profits and the NRL it would be 6.1 (basic research), for the in-house centers it would be 6.2 (applied research), and for industry it would be 6.3 (advanced technology development).

As mentioned, attracting the necessary human resources requires hiring about 500 new S&Es (preferably with Ph.D.s) annually, and providing incentives to train and retain them. One of the best ways to attract technical talent is simply to offer the opportunity and the means to carry out first class research. That is, promising new hires a stable funding source during their transition from academia to the government, and also access to the best facilities and equipment, would create a strong incentive for them to work for the DON.

The tri-service NRAC report *Science and Technology Community in Crisis*, which looked at the corporate laboratories owned and operated by the three Services, discussed this issue at length. The study panel visited each of the Service's research laboratories and interviewed a cross section of their new hires, most of whom had Ph.D.s. The panel remarked:

One of the issues that came up...was that of funding, especially for new or recent hires. Lack of stable funding for new hires, especially those engaged in basic research, can impede efforts to recruit. Part of the problem is that many managers (who are key to recruiting new employees) worry that they will be unable to provide funding for new hires. This makes them cautious about hiring, and so less willing to actively recruit.

During its visits, the Panel also heard concerns from some of the new hires over pressure to obtain funding for themselves. Turnover is high in the first few years of employment under any circumstances and can be exacerbated by the lack of stable funding. Therefore, the Panel considers that there may be value in establishing mechanisms for ensuring (at least) partial support for new S&E hires, especially those working in basic research. This would relieve some pressure on new hires to obtain their own funding, and create an environment more attuned to hiring and retention at the entry level.[15]

The ability of TDs to provide new hires with a stable source of funding, at least for the first two years, would indeed be a strong recruitment and retention tool. It would also allow new hires to focus only on their research. This in turn would both ease their transition and expedite the rate at which they acquire the skills and knowledge to reach the full performance level. Last but not least, it would reduce the heavy attrition in the first few years after hiring.

Recommendation 6: Provide a $50 million laboratory and equipment-funding source in the DON S&T account to be focused on S&T frontiers.

For reasons outlined in detail in Chapter 4, it is becoming nearly impossible to acquire the facilities and equipment that could attract potential new members of the S&T workforce. Indeed, such tools are the fundamental element of world-class research itself. In a rapidly evolving global technology base, where equipment can become obsolete almost overnight, even an outstanding researcher cannot do his best work without state of the art facilities. Considering that many current DON in-house facilities are unique, they must be furnished with adequate funding for necessary upgrades and enhancements.

There are many possible ways to address the problem. Table 6.2 suggests three options but there are certainly others. Option 1 would restrict the competitive field for the laboratories and centers to a group consisting only of RDT&E-related MILCON projects. They would not have to compete head-to-head with readiness-related projects. Option 2 would give the laboratories and centers a vote in the CNI/N4 MILCON ranking process. Option 3 would essentially give them a champion for their MILCON needs in the form of the Deputy Chief of Naval Operations for Warfare Requirements and Programs (N6/N7). Encouraged by the ASN (RDA), this office would argue for facility needs during the larger prioritization process.

Table 6.2: Some Options for Addressing NLCCG Community MILCON Needs

Option 1	NLCCG MILCON compete in a "fenced" CNO RDT&E programmatic category with the resulting Integrated Priority List endorsed by the ASN (RDA)
Option 2	CNI give mission claimants a MILCON vote in the new ranking process
Option 3	ASN (RDA) pursue more N6/N7 sponsorship for RDT&E projects

Why should DOD's in-house researchers be deprived of essential tools when their counterparts have many avenues for acquiring facilities and equipment and seeking funds from parent institutions? The NSF, for example, provides academia such funds through its Major Research Equipment and Facilities Construction (MREFC) account. Its rationale for doing so is succinctly summarized in its 2005 facility plan:

Successful exploration—whether in uncharted wilderness or at the frontiers of knowledge—demands commitment, vision, daring and ingenuity. But those qualities alone are not always sufficient. Progress also requires the right kind of equipment. Often new territory is accessible only with new tools; and sometimes even a seemingly unstoppable rush of discovery must halt to await novel means of seeing, manipulating and analyzing natural phenomena.... That is why the [NSF] supports not only research and education, but also the physical implements that make both possible.[16]

The MREFC budget line supports investments that range from modest laboratory instruments and information technology resources to world-class projects. Its budget in FY 2004 was $173.7 million, and the Bush Administration requested an FY 2005 budget of $250 million. FY 2004 projects ranged in size from $8 million to $50.7 million, enabling construction of several very large facilities.

The DOD itself also provides funding to academic institutions through the Defense University Research Instrumentation Program (DURIP).[17] Part of DOD's University Research Initiative, this program helps colleges and universities improve their research capabilities and educate S&Es in areas important to national defense. Funds can be used to acquire major equipment to augment capabilities or develop new ones. Proposals for equipment purchase—which can range from $50,000 to $1,000,000—are evaluated competitively. Overall, for FY 2006 the DOD intends to award approximately $41 million via grants from the Services' respective research offices. In FY 2005, 212 awards totaled $43.9 million, and individual awards ranged from about $60,000 to $990,000, with an average of $207,000.

The DON centers should also have access to a funding line for laboratories and scientific equipment. It is therefore proposed that the DON create a $50 million funding source in the S&T account to be focused on new frontiers. The S&T Governance Council, recommended earlier in this chapter, could oversee the annual proposal process.

6.5   Inter-Generational Transfer of Knowledge

Recommendation 7: Launch an aggressive DON-wide program to ensure inter-generational transfer of corporate knowledge.

Today, both the DOD centers and many American companies are failing to transfer critical knowledge from older employees approaching retirement. One recent study found that few organizations are even capable of doing this. Based on a survey of more than 500 full-time U.S. workers between 40 and 50 years of age, the study discovered that 45 per cent have no formal planning processes or tools to capture their knowledge. Moreover, a quarter said their employers will let them retire without any prescribed transfer. Only one in five said they anticipated an intensive, months-long method of knowledge transfer, while only 28 per cent said they expected a formal process lasting one or two weeks. A

158

further 16 percent anticipated only some informal discussion with others prior to retirement. The study also found that few companies take advantage of the experience and expertise of their retired workforce.[18]

Given that more than 25 percent of the working U.S. population will reach retirement age by 2010, these companies must act soon or face a major exodus of institutional knowledge. The study recommended three critical steps: First, companies *"must understand the extent of the problem, including the skills at risk, and their organization's ability to tackle it."* Second, they need *"a strategy to capture and transfer core skills from retiring employees and to identify, attract and retain new workers with critical skills."* And third, they must proactively manage the effort. *"The bottom line is that leaders in this arena know that capturing critical workforce knowledge and skills can't be left to chance."*

As discussed at length in Chapter 4, the impending retirement of the Baby Boom S&Es in the DOD coupled with the dwindling pool of potential new hires is creating this same problem of knowledge management, a term that applies to practices that create, organize and leverage collective knowledge to enhance organizational performance. Generally, there are three forms of organizational knowledge: *tacit* (held by people), *explicit* (codified in documents and databases), and *social* (personal networks). As with the related issue of corporate memory, nothing has been done on a broad scale in the DOD to facilitate the spread of this knowledge, meaning it will be nearly impossible for laboratories and centers to build new core competencies or even maintain their current ones.[19]

In fact, evidence suggests they are not even doing the job they once did to preserve their explicit knowledge. For example, an ONR text mining effort looked at how many publications the DON technical community deposited with the DTIC during three time intervals. Table 6.3 records the dramatic drop off in the DTIC database. While there could be multiple reasons for the decline, they are all likely related to the recent, significant loss of intellectual capital experienced as a result of personnel reductions imposed by cost-savings initiatives and several BRAC actions. Regardless of the cause, the result is a failure to record much of the community's important explicit knowledge.

While explicit knowledge is key, much of the important organizational knowledge in the DOD laboratories and centers is tacit in nature. Once this knowledge base is gone, reconstitution may not be achievable, or it may take years. It is important that the DON act now to ensure it is passed along to new employees before the sand runs out of the hourglass.

Another simple but potentially effective tool is an exit questionnaire, a record of key contacts and learning stories, processes, and practices. Most of the important tacit knowledge, however, is best transferred through formally established programs. In this regard, mentoring could be an explicit element of the job description for key carriers of corporate knowledge. For those highly knowledgeable S&T workers nearing retirement, it could even be a full-time effort. Regardless of the strategies, they must be formally planned, and underpinned by necessary financial resources. Only in this way can experienced employees teach new ones the explicit and tacit knowledge that constitute the organization's Deep Smarts.

DON Activity	Year			
	1980 - 1985	1990 - 1995	2000 - 2005	Total
NSWC	3,203	2,953	1,195	30,827
NAWC	3,601	2,837	871	36,877
NUWC	1,497	587	282	7,424
SSC	1,775	3,025	578	17,759
Warfare Center Subtotal	10,076	9,402	2,926	92,887
NRL	2,661	3,000	1,279	20,171
DON Lab/Center Subtotal	12,737	12,402	4,205	113,058
U.S. Naval Academy	87	106	87	637
Naval Post Graduate School	3,478	5,663	3,842	25,164
DON Total	16,302	18,171	8,134	138,859

Table 6.3: DON Technical Publications in DTIC Database

The issue is especially acute in the world of research, where much of the knowledge is retained in the individual. A few private sector companies have addressed this, realizing that one of the best ways to pass along corporate knowledge is to have mentors working closely with young researchers. It has long been recognized that transitioning scientific and technological information is a "contact sport," and the gray beards must be brought together with the new people in S&T to ensure corporate information that has been obtained, many times at great expense, is passed along. Indeed, it is more cost effective to preserve now than to try to reconstitute later.

As just one example, consider the research area of EMs, where the S&E workforce is shrinking dangerously. In such areas, where the DOD must retain in-house technical capability, it is imperative to act now. In-house mentoring efforts and collaborative arrangements with private sector partners such as the CECD are crucial. Also, innovative hiring measures could be used to bring back key personnel after they retire, specifically to facilitate knowledge transfer.

6.6 A New S&T Enterprise Requires Visionary Leadership

Recommendation 8: Create an S&T Governance Council chaired by the ASN (RDA) with membership that includes all major stakeholders, including the ONR, the NRL, the Warfare/Systems Centers and UARCs.

The creation and successful operation of a new DON S&T enterprise will require the sustained leadership of individuals at the headquarters level of both the OSD and the Services, and at the laboratory/center TD level as well. This leadership must provide both a compelling vision of what this new enterprise should look like as well as the guidance and inspiration to translate it into reality. This undertaking will not be an easy one, nor will the final result be achieved swiftly. Indeed, few worthwhile transformation efforts are achieved quickly or easily. Consequently, if it is to be successful, the leadership that guides it must remain in place long enough to provide the necessary follow-through on implementation plans. In the past, this lack of leadership continuity has been a significant impediment to change because the often short tenure of senior leaders—civilian and military—failed to provide the consistent, unqualified support and advocacy needed to achieve it.

In recent years the Secretary of Defense has said little about the importance of in-house S&T, and instead, that task has been largely left to DDR&E, a situation unsuited for consistent, high-level, advocacy within overall budget priorities. As mentioned in Chapter 4, lack of strong DDR&E leadership has concerned many in the Congress and prompted a Section (911) in the FY 2005 National Defense Authorization Act that called for a study of that office. In the report that accompanied the authorization bill, the Senate Armed Services Committee expressed concern *"about the gradual deterioration of the authority and stature of the ODDRE over the last decade. In the past, this position, or its equivalent, played the leading role in planning, programming, and coordinating the research and technology programs of the Department. Historically, [it] also had considerable budgetary authority to invest in critical defense technologies and significant access to the highest levels of defense leadership. More recently...the DDRE plays more of a consultation and liaison role with the services and S&T components and has struggled to raise the profile of S&T priorities...especially within acquisition program offices and the warfighting community. As the Department continues to pursue transformational capabilities, which are heavily dependent on technology, a strong S&T advocate is critical.*[20] [Emphasis added]

The intent of this recommendation is not to replace the mission and functions of the ONR, but merely to provide the comprehensive Secretariat-level attention sorely lacking for the laboratory/center community. The DON especially needs such a champion for S&T investments with longer-term payoffs. In a budget-constrained environment, this is vital to help the D&I portion of the investment "compete" with other claimants for scarce resources by providing the technical community comparable access to senior decision-makers.

Recommendation 9: Establish SES-level TD positions at the warfare/systems center division sites, and invest them with authority and responsibility for the entire technical output of the organization.

Most analysts agree on the elements of effective leadership in an R&D organization. The 1991 Adolph Commission, for example, argued that an essential component of a world-class technical institution is "an inspired, empowered, highly qualified leadership committed to technical excellence through support for excellence, creativity, and high-risk/high-payoff initiatives."[21] Effective leaders must know how to balance long-range challenges against immediate technology needs, and costs to benefits. They must also be adept at choosing the technical areas in which to work, divesting work more appropriate for other performers (transition), judging scientific and technical merit, and orchestrating the conditions that foster innovation. Given the long horizon for much of the work, leadership continuity is also essential.

Because the TD position proved especially important to cutting edge technical institutions, credentials for the job have been widely discussed. Qualifications enumerated include not only a strong technical background but also technology management experience; a commitment to a creative work environment that encourages individual initiative; a long-term perspective of planning, accomplishment, and resource commitment, because breakthroughs can take many years to mature into operational applications; and a willingness to undertake work recognized as being high risk and having high payoff potential. It is worth noting in this regard that the DOE appoints only internationally recognized scientists and engineers to head its national laboratories.

In addition to the TD's authority, overall technical leadership at the warfare/system centers has been allowed to deteriorate as well. Consider for example NSWC's Carderock, Maryland division. Ten years ago, the civilian leadership consisted of 13 Senior Executive Service (SES) level individuals: the TD, six technical department heads, two senior staff positions (Director of Strategic Planning and S&T Director), and four senior technical (ST) positions. Today, the division has two SES department heads, one Senior Scientific Technical Manager (who serves as the technical operations manager) and four STs.

To remedy this situation, the DON should formally re-establish the dual executive management approach. Importantly, it should reinstitute the position of TD at each warfare/system center division site and ensure the position has responsibility and authority for the entire technical program. Given its importance and responsibilities, the TD job should be an SES position. Moreover, the incumbent should be of outstanding ability and achievement, and have the respect of peers throughout the technical community.

The position of TD is essential because, regardless of the competency of commanding officers, they do not have the same bond with either the scientific community at large or the S&Es who work for them. Nor does their rotation allow for continuity of technical oversight. In fact, many laboratory/center commanders have expressed this view. For example, VADM William J. Moran, USN retired, asserted:

I'm a real believer in a strong Technical Director with competence and knowledge of the Navy's problems and, hopefully, if not an aggressive personality, at least an affirmative personality....

One [reason] is that any effort...is going to take so many more years to completion than it used to that someone in the laboratory in a senior position...has to be there to provide the continuity and the memory from the beginning to the end of a program.... If we had...a large group of uniformed officers who had operational experience and knowledge of the operating problems and advanced technical education and technical depth and had grown up working in the material side of the house or the laboratory system...if we had a group of officers who fit that description, I'd be much more enthusiastic about the commanding officer of a laboratory having a strong voice in the technical programs. But, the kind of background I just described would take you 40 years to get. Now, we just can't get there....

We have, in the record, any number of cases where the new CO comes aboard and he decides everything that was done in the last two years or three years was done incorrectly, or shouldn't be done at all, and you get major reorientations and wrenching of the program.[22]

Admiral Moran, like many other naval officers, recognizes the necessity of oversight continuity in long-term S&T efforts. Given this need, all appointments to the TD position must be at least five years in duration.

6.7 Military Officers as an Asset for the S&T Community

Recommendation 10 Institute within military career paths a cadre of "Military Technology Officers".

As noted, DOD laboratories and centers have benefited greatly from having young military officers with technical backgrounds working along side civilian S&Es. Civilians gained a greater appreciation of the warfighters' perspective while the officers gained appreciation of how S&T develops future military capability. As many studies have observed, however, the number of young officers in the community has dwindled.

The Air Force, where the decline was especially noticeable, has acted decisively to reverse it.[23] When the Air Force was created in 1947, Dr. Theodore Von Karmen set it on a course where officer knowledge of technology was essential. That attitude began to atrophy in the mid-1980s and reached bottom in the early 1990s, when senior leadership made it clear that operational experience was the best path to promotion. However, there has been a turnabout. For example, enrollments at the AFIT have nearly tripled since the late 1990s when the Secretary of the Air Force tried to close it. There are now some 400 officers in M.S. engineering programs there and 35 in Ph.D. programs. The budget suggests higher numbers in the out years, due largely to a special initiative by former Air Force Secretary Roach. Most of these officers go either to the Air Force Research Lab (AFRL) or to program special project offices, and the ratio of officers to civilians in AFRL is slowly climbing. Although hard data are elusive, some evidence suggests that promotions for Ph.D. and M.S. degree S&E officers are now on a par with the remainder of the Air Force line officers. Further, the Service now sends almost 1,000 majors and lieutenant colonels to command and staff school at Maxwell Air Force Base, the mid-level of professional military education. Some 100 from each class (regardless of academic background) are sent to AFIT for technical courses for one year in addition to the policy doctrine courses. The hope is that they will obtain and appreciate technical knowledge as they assume senior staff and command positions.

There is some effort in the Navy along similar lines. Very few young officers have advanced technical degrees, nor even an appreciation for the role of S&T in providing warfighting capability. This is one reason N-STAR cultivates relationships and continuous interactions between U.S. Naval Academy (USNA) midshipmen and young S&Es throughout the DON laboratory/center community. For example, at the inaugural N-STAR Conference at the USNA in September of 2005, over 70 guest speakers from the warfare centers lectured and conducted technical sessions with USNA professors and Trident Scholars on topics related to emerging S&T. Representatives from throughout the Naval Research Enterprise gave hands-on previews of what is in store for the midshipmen, including demonstrations of cutting-edge technology under development.

Even so, much more needs to be done. Examples could include an internship program for midshipmen that included at least one summer tour at a DON laboratory or center, or a broader USNA/Naval Postgraduate School (NPS) summer faculty program wherein the schools collaborate on research projects with laboratory/center personnel. Besides the obvious advantage of carrying out meaningful research, there would be other benefits from such arrangements. For example, faculty who work in the technical community would likely bring success stories back to the students, conveying a better understanding of the applications of science and engineering to military problems. Such communication and confidence-building efforts would provide a significant BtC ratio.

Because high technology organizations like the ONR, NRL, and the warfare/systems centers need a cadre of technically sophisticated officers to work with civilian S&Es in line organizations, the DON should create within appropriate military career paths a cadre of "Military Technology Officers." There is a particular need for more officers with Ph.D. degrees. As in the case of the Air Force, these changes would require that senior leadership recognize the importance of this career path and ensure that those who follow it are appropriately promoted.

Officers who have retired or mustered out of service form another valuable reservoir of talent the laboratories and centers could tap to help fill the S&T hiring pipeline. In fact, one component of N-STAR seeks to send former officers back to school for advanced technical degrees. In return for having their degree funded, they would be required to work at a DON laboratory or center. In FY 2006 the first of this pool will start a Ph.D. program at the NPS, where the candidate will have both an advisor at NPS and a mentor at one of the DON's laboratories or centers. This is but one example of how such talent can be recycled, and of how such initiatives must be underwritten with necessary resources.

6.8   Final Thoughts

Despite a track record of success in planting the seeds for future military capabilities, the DOD's in-house S&T effort continues to reel from nearly 15 years of downsizing, consolidation, closure, and outsourcing. Reversing the decline will not be easy, as this book has demonstrated. In fact, the growing shortage of young S&Es and rapid globalization threaten not only the defense S&T effort, but also America's economic and national security.

Recently, the Congress asked the National Academies how to meet these challenges. Their answer came in the form of a major report called *Rising Above the Gathering Storm: Energizing and Employing America for a Brighter Economic Future.* Principal among their recommendations were the following:[24]

- Increase America's pool of technical talent through K-12 education initiatives
- Sustain and strengthen America's commitment to long-term basic research
- Make American the most attractive setting in which to research, thereby making it an attractor of the world's best talent
- Provide incentives to create and sustain an innovation environment

The Academies' proposals are aligned with many of those made here, and indeed, it is hoped this book will help the DON rise above its own approaching storm, and thereby help America remain secure and prosperous.

The recommendations here focus on revitalizing the DON's in-house laboratories and centers, in particular the S&T effort. As has been noted, may of the suggestions are not unique, nor the list provided intended to be complete. Indeed, a major purpose of this book is to stimulate a dialogue and elicit other constructive steps. However, the roadmap proffered here would put the DON S&T enterprise on a good path to the future. While the benefits of revitalizing the in-house S&T effort would indeed be enormous, and would outweigh any costs, the consequences of failing to ensure its vitality could well be catastrophic.

In the early days of World War II, Hitler's U-Boats wreaked havoc on Allied merchant and military ships. Fortunately, allied scientists developed an answer—SONAR. But sonar was no overnight invention. It was the accumulated result of years of research by scientific prospectors on both sides of the Atlantic. As Norman Cousins, the well-known editor and essayist, observed, *"History is a vast early warning system."*[25] In this sense, sonar aptly represents the entire issue: like sonar, and like Cousins' history, the S&T enterprise is a kind of early warning system, which can not only prevent but also defeat aggression by insuring technological preparation for 21st-century challenges.

Notes

Chapter 1

[1] U.S. Commission on National Security, "Phase I Report: The Emerging Global Security Environment for the First Quarter of the 21st Century," September 15, 1999.

[2] Michael S. Teitelbaum, "Do We Need More Scientists?", *The Public Interest*, No. 153 (Fall 2003).

[3] Shirley Ann Jackson, "Envisioning a 21st-Century Science and Engineering Workforce for the United States," Report to the Government-University-Industry Research Roundtable of the National Academies, 2003.

[4] Shirley Ann Jackson, "Sustaining our National Capacity for Discovery and Innovation," Presentation to Business Roundtable, June 16, 2004.

[5] William P. Butz, *et al.*, "Is There a Shortage of Scientists and Engineers? How Would We Know?", RAND Science and Technology Issue Paper, November 12, 2002.

[6] Jeffrey Mervis, "Down for the Count?", *Science*, Vol. 300, May 16, 2003.

[7] Mary Anne Fox, "Meeting Summary: Pan-Organizational Summit on the Science and Engineering Workforce," Government-Industry-University Research Roundtable, November 11-12, 2002, The National Academies, Washington, D.C.

[8] "Industry, DOD Strategize to Avert Workforce Crisis," *InsideDefense.com*, Defense Alert, December 23, 2004.

[9] Leonard Wiener, "Brain Drain," *U.S. News & World Report*, November 22, 2004.

[10] Sandra I. Erwin, "Technical Skills Shortage Hurts Pentagon Bottom Line," *National Defense*, September 2004, accessed at

http://www.nationaldefensemagazine.org/issues/2004/Sep/Technical_Ski lls.htm.

[11] General Accounting Office, "Military Bases: Review of DOD's 1998 Report on Base Realignment and Closure," GAO/NSIAD-99-17, November 9, 1998, accessed at http://www.gao.gov.

[12] Michael L. Marshall, "Defense Laboratories and Military Capability: Headed for a BRACdown?" *Defense Horizons* 44, Washington, D.C.: National Defense University Press, July 2004, accessed at http://www.ndu.edu/ctnsp/publications.html.

[13] Don J. DeYoung, "The Silence of the Laboratories," *Defense Horizons* 21, Washington, D.C.: National Defense University Press, January 2003, accessed at http://www.ndu.edu/ctnsp/publications.html.

Chapter 2

[1] Director of Defense Research and Engineering Memorandum for the Assistant to the President for Science and Technology, Subject: Department of Defense Interim Response to NSTC/PRD #1, Presidential Review Directive on an Interagency Review of Federal Laboratories, September 12, 1994.

[2] Department of Defense, "Federally Funded Research and Development Centers (FFRDC) Management Plan," May 1, 2000.

[3] Director of Defense Research and Engineering Memorandum for the Assistant to the President for Science and Technology, Subject: Department of Defense Response to NSTC/PRD #1, Presidential Review Directive on an Interagency Review of Federal Laboratories, February 24, 1995.

[4] Director of Defense Research and Engineering Memorandum, September 12, 1994, *op. cit.*

[5] James E. Colvard, "Closing the Science-Sailor Gap," U.S. Naval Institute *Proceedings*, June 2002.

[6] Secretary of Defense Memorandum of October 14, 1961, Subject: In-House Laboratories. This memorandum is an exhibit in "Federal Budgeting for Research and Development: Hearings before the Subcommittee on Reorganization and International Organizations of the Committee on Government Operations United States Senate," U.S. Government Printing Office, 1961. This is a compilation of material related to hearings held before the subcommittee July 26-27, 1961.

[7] ASN (Research and Development), Naval Research Advisory Committee, Ad Hoc Subcommittee for the Study of Navy Laboratory Utilization (August 1974), John Allen, Rodney Grantham, and Donald Nichols, "The Department of Defense Laboratory Utilization Study," April 28, 1975.

[8] Executive Office of the President, Office of Science and Technology Policy, "Interagency Federal Laboratory Review Final Report," May 15, 1995.

[9] The plan for the creation of the DON laboratory/center community, "RDT&E Engineering and Fleet Support Activities Consolidation Plan," was approved by the Secretary of the Navy and forwarded to Navy components for implementation under cover letter dated April 12, 1991.

[10] Albert B. Christman, Robert M. Glen, and Richard E. Kistler, "Evolution of Navy Research and Development," Director of Navy Laboratories, Washington, D.C., September 30, 1980.

[11] David K. Allison, Historian of Navy Laboratories, "Evolution of Missions and Functions of CNM Commanded Laboratories and Centers," November, 1981.

[12] Rodney Carlisle, "Management of the U.S. Navy Centers during the Cold War: A survey guide to reports," Navy Laboratory/Center Coordinating Group and the Naval Historical Center, Washington, D.C., 1996.

[13] Rodney Carlisle, "Navy RDT&E Planning in an Age of Transition: A Survey Guide to Contemporary Literature," Navy Laboratory/Center Coordinating Group and the Naval Historical Center, Washington, D.C., 1997.

[14] The establishment of these two groups was formalized in Secretary of the Navy Memorandum, 5400.16 of December 18, 1992.

[15] On November 29, 2002, the ASN(RDA) signed out a terms of reference for the NLCCG which, inter alia, expanded the membership of the NLCCG to include the Deputy Assistant Secretary of the Navy for Research, Development, Test and Evaluation, as well as the commanding general and technical director of the Marine Corps Warfighting Laboratory.

[16] These and other changes were promulgated by the ASN(RDA) in a memorandum of November 29, 2002, Subject: Navy Laboratory/Center Coordinating Group.

[17] The overall Federal budget is broken down into several budget titles or functions. National Defense (051) consists of several subtitles: DOD funding (051), defense-related spending in the Department of Energy (053) and defense-related spending in other Federal departments (054). Subtitle 051 itself consists of a number of funding lines: RDT&E, Operations and Maintenance, Procurement, Military Construction, Family Housing, and funding for Military Personnel.

[18] Testimony of Dov Zakheim before the House Armed Services Committee, June 17, 1997.

[19] John P. White and John Deutch, "Building Capability from the Technical Revolution That Has Happened: Report of the Belfer Center

Conference on National Security Transformation," June 2004, accessed at http://www.carlisle.army.mil/ssi/pdffiles/PUB407.pdf.

[20] This Circular provides Federal policy governing the contracting out of so-called commercial activities. The most recent version, dated May 29, 2003, can be accessed at http://www.whitehouse.gov/omb/circulars/a076/a76_incl_tech_correction.pdf.

[21] Hiring freezes can have a significant impact on laboratory demographics. A discussion of this can be found in William J. Ferreira, "Demographic Consequences of Continuing Constraints on Hiring: A Technical Institutions Perspective," Naval Surface Warfare Center, Dahlgren Division, May 1993.

[22] Private Communication, Ms. Betty Duffield, Director, Human Resources Office, Naval Research Laboratory, July 1, 2002.

[23] Don J. DeYoung, "The Silence of the Labs," *Defense Horizons* 21, Washington, D.C.: National Defense University Press, January 2003, accessed at http://www.ndu.edu/inss/DefHor/DH21/DH21.pdf.

[24] As an example of this recent direction, the Commander of NAVSA, in an email of January 27, 2005, had this to say about workforce shaping at NSWC's Crane Indiana Division: "*Warfare Center hiring strategies must be executed in accordance with an attrition-based, right-sizing approach with a focus on efficiency ... Consider a more aggressive workforce shaping target to achieve an end-of-FY05 actual below your plan of 2,660.*"

[25] Michael L. Marshall and J. Eric Hazell, "Private Sector Downsizing: Implications for DOD," *Acquisition Review Quarterly*, Spring 2000, accessed at http://www.findarticles.com/p/articles/mi_m0JZX/is_2_7/ai_78178118.

[26] National Academy of Public Administration, "Civilian Workforce 2020: Strategies for Modernizing Human Resources Management in the Department of the Navy," (NAPA 2020 Study), August 18, 2000.

[27] Statement of Sean O'Keefe, Administrator, National Aeronautics and Space Administration, before the Subcommittee on Space and Aeronautics, Committee on Science, House of Representatives, July 19, 2002, accessed at http://www.house.gov/science/hearings/space02/jul18/okeefe.htm.

[28] Participating laboratories included Argonne National Laboratory, Brookhaven National Laboratory, Fermi National Accelerator Laboratory, Idaho National Environmental Engineering Laboratory, Lawrence Livermore National Laboratory, Los Alamos National Laboratory, National Energy Technology Laboratory, National Renewable Energy Laboratory, Oak Ridge National Laboratory, Pacific Northwest National Laboratory, Princeton Plasma Physics Laboratory,

Stanford Linear Accelerator Laboratory, Sandia National Laboratories, and the Savannah River Technology Center.

[29] Private communication of June 6, 2001 with Sheryl Hingorani of Sandia National Laboratory.

[30] Naval Research Advisory Committee, "Science and Technology Community in Crisis," May 2002, accessed at http://nrac.onr.navy.mil/webspace/reports/S-n-T_crisis.pdf/.

[31] Similar concerns about Air Force laboratories were voiced in "Effectiveness of Air Force Science and Technology Program Changes," a study mandated by Congress under Section 253 of the FY 2002 National Defense Authorization Act. Accessed at http://www.nap.edu/openbook/030908895X/html.

[32] Committee on the Future of the U.S. Aerospace Infrastructure and Aerospace Engineering Disciplines to Meet the Needs of the Air Force and the Department of Defense, Air Force Science and Technology Board, Division on Engineering and Physical Sciences, National Research Council, "Review of the Future of the U.S. Aerospace Infrastructure and Aerospace Engineering Disciplines to Meet the Needs of the Air Force and the Department of Defense," Washington D.C.: The National Academies Press, 2001, accessed at http://www.nap.edu/books/0309076064/html.

[33] Commission on the Future of the United States Aerospace Industry, "Final Report of the Commission on the Future of the United States Aerospace Industry," November 18, 2002, accessed at http://ita.doc.gov/td/aerospace/aerospacecommission/AeroCommissionFinalReport.pdf.

[34] See, for example, the July 22, 2004 statement of John W. Douglass, President and Chief Executive Officer of the AIA before the Strategic Forces Subcommittee of the House Armed Services Committee, accessed at http://www.house.gov/hasc/openingstatementsandpressreleases/108thcongress/04-07-22Douglass.pdf.

[35] National Defense Industrial Association, "Industry Position on Critical Workforce Skills," Quick-Look Report, June 10, 2004.

[36] A perfect storm has been defined as a situation where, by the confluence of specific events, what might have been a minor issue ends up being magnified to proportions that are out of control.

[37] Robert A. Frosch, "The Customer for Research and Development is Always Wrong," *Research Technology Management*, Vol. 39, No. 6, November/December 1996.

[38] Bureau of the Budget, "Report to the President on Government Contracting for Research and Development," (Bell Report), April 30, 1962.

[39] *Ibid.*

[40] Report on the Audit of In-House Laboratory Independent Research Program, Project 5AB-060, February 1986.

[41] Secretary of Defense Memorandum of October 14, 1961 to the Secretaries of the Military Departments, the Director of Defense Research and Engineering, and the Assistant Secretary of Defense (Manpower), Subject: In-House Laboratories. This memorandum is an exhibit in "Federal Budgeting for Research and Development: Hearings before the Subcommittee on Reorganization and International Organizations of the Committee on Government Operations United States Senate," U.S. Government Printing Office, 1961. This is compilation of material related to hearings held before the subcommittee, July 26-27, 1961.

[42] Cited in David Jewell, "Independent Research and Independent Exploratory Development in RDT&E Centers Commanded by the Chief of Naval Material," December 31, 1981. Jewell was Special Assistant to the Director of Laboratory Programs in the Naval Material Command.

[43] Department of Defense Instruction Number 3201.4 of October 8, 1993, Subject: In-House Laboratory Independent Research (ILIR) and Independent Exploratory Development (IED) Programs, accessed at http://www.dtic.mil/whs/directives/corres/html/32014.htm.

[44] Office of Science and Technology Policy, Report of the White House Science Council Federal Laboratory Review Panel, May 1983

[45] U.S. Department of Defense, Office of the DDR&E, Report of the Defense Science Board, "1987 Summer Study on Technology Base Management," Springfield, VA, December 1987.

[46] Office of Technology Assessment, "Holding the Edge: Maintaining the Defense Technology Base," April 1989, accessed at http://www.wws.princeton.edu/cgi-bin/bytesery.prl/~ota/disk1/1989/8920/8920.PDF.

[47] Hugh Montgomery, "DOD Science & Technology Invigoration," Potomac Institute for Policy Studies' Report: PIPS-02-01, February 2002, accessed at http://www.potomacinstitute.org/publications/studies/dodsandtint.pdf.

[48] Private communication with Steve Persaud, Naval Surface Warfare Center, Dahlgren Division, May 17, 2005.

[49] Sandra I. Erwin, "Technical Skills Shortage Hurts Pentagon's Bottom Line," *National Defense*, September 2004, accessed at http://www.nationaldefensemagazine.org/issues/2004/Sep/Technical_Skills.htm.

[50] Colvard, "Closing the Science-Sailor Gap," *op. cit.*

[51] White House Report: Federal Coordinating Council for Science, Engineering and Technology Report, "Applications of OMB Circular A-76: An R&D Management Approach," October 31, 1979.

[52] Albert B. Christman, Robert M. Glen and Richard E. Kistler, *op. cit.*

[53] Federal Advisory Commission on Consolidation and Conversion of Defense Research and Development Laboratories, Report to the Secretary of Defense, September 1991. The Commission was established by Public Law 101-510 to study the DOD laboratories and provide recommendations to the Secretary of Defense on the feasibility and desirability of various means to improve the operation of the DOD laboratories. It is often referred to as the "Adolph Commission," after its chairman, Charles Adolph.

[54] Director of Defense Research and Engineering Memorandum for the Assistant to the President for Science and Technology, Subject: Department of Defense Interim Response to NSTC/PRD #1, Presidential Review Directive on an Interagency Review of Federal Laboratories, September 12, 1994.

[55] James Colvard, "Savings Can Have a High Price," *Government Executive Magazine*, November 1, 1998, accessed at http://www.govexec.com/features/1198/1198defense.htm.

[56] "Boeing 'Brain Drain' Linked to Columbia Accident," *Spacetoday.net*, July 31, 2003, accessed at http://spacetoday.net/sts107.php?sid=1835&type=s.

[57] Erwin, *op. cit.*

[58] Ann R. Markusen, "The Case Against Privatizing National Security," *Governance: An International Journal of Policy, Administration, and Institutions*, Vol. 16, No. 4, October 2003.

[59] Michael L. Marshall and J. Eric Hazell, "Outsourcing Research and Development: Panacea or Pipe Dream?" Naval Institute *Proceedings*, October 2000.

[60] Erwin, *op. cit.*

[61] J.R. Harbison, T.S. Moorman, Jr., M.W. Jones, and J. Kim, "U.S. Defense Industry Under Siege—An Agenda for Change," Booz-Allen Hamilton report, July 2000.

[62] Figures based on Security Exchange Commission 10K filings of Boeing, General Dynamics, Lockheed Martin, Litton, Northrup Grumman, Raytheon and TRW.

[63] Hugh Montgomery, *op. cit.*

[64] Amy Klamper, "Pentagon Budget Cuts Might Squeeze Firms," *Congress Daily*, February 4, 2005, accessed at http://www.govexec.com/dailyfed/0105/010405cdam1.htm.

[65] Roxana Tiron, "Future Pentagon Investments To Reshape Defense Industry," *National Defense*, National Defense Industrial Association,

May 2005, accessed at http://nationaldefense.ndia.org/issues/2005/May/Future_Pentagon.htm.

66 National Science Board, "Science and Engineering Indicators-2004", Table 4-1, accessed at http://www.nsf.gov/sbe/srs/seind04.

67 Eugene B. Skolnikoff, "Research Universities and National Security: Can Traditional Values Survive," supplement to the American Association for the Advancement of Science, Science and Technology Policy Yearbook 2003, Albert H. Teich, Stephen D. Nelson, and Stephen J. Lita, editors, accessed at http://www.aaas.org/spp/yearbook/2003/yrbk03a.htm.

68 Genevieve J. Knezo, "Possible Impacts of Major Counter Terrorism Security Actions on Research, Development, and Higher Education," Congressional Research Service Report for Congress, April 8, 2002, accessed at http://www.aau.edu/research/crsterror.pdf.

69 Michael L. Marshall, Timothy Coffey, Fred E. Saalfeld, and Rita R. Colwell, "The Science and Engineering Workforce and National Security," Defense Horizons 39, Washington, D.C.: National Defense University Press, April 2004, accessed at http://www.ndu.edu/ctnsp/defense_horizons/DH39.pdf.

70 Ron Southwick, "Pentagon Considers Tighter Controls on Academic Research," Chronicle of Higher Education, April 24, 2002; "Defense Department Proposes New Restrictions on Research Data," Inside the Pentagon, May 2, 2002; Mary Leonard, "Pentagon Drops Attempt to Keep Research Quiet," Boston Globe, May 9, 2002, A3; and Don J. DeYoung, "Proposed Security Controls on Defense Research," April 2, 2002, accessed at http://www.fas.org/sgp/othergov/deyoung/html.

71 David Malakoff, "Pentagon Proposal Worries Researchers," Science, May 3, 2002.

72 In its most general sense, globalization refers to changes in societies and the world economy resulting from a rapid escalation in cultural exchanges and trade. Discussions of the trade aspect typically center on the growing linkages between products, markets, firms, and production factors, with an increasing proportion of each derived, generated, or available more widely among the various geographic areas of the world. The Internet and other information technology sources have greatly accelerated the process of globalization.

73 See, for example, congressional concerns voiced in "Offshore Outsourcing and America's Competitive Edge: Losing Out in the High Technology R&D and Services Sectors," Office of Senator Joseph I. Lieberman, May 11, 2004, accessed at http://lieberman.senate.gov/newsroom/whitepapers/Offshoring.pdf.

74 Michael L. Marshall, Timothy Coffey, Fred E. Saalfeld, and Rita R. Colwell, op. cit.

[75] Dana Hicks, "Asian Countries Strengthen Their Research," *Issues in Science and Technology*, Summer 2004, accessed at http://www.issues.org/issues/20.4/realnumbers.html.

[76] Task Force on the Future of American Innovation, "The Knowledge Economy: Is The U.S. Losing Its Competitive Edge?", February 16, 2005, accessed at http://www.futureofinnovation.org/PDF/Benchmarks.pdf.

[77] Ibid

[78] Francis Narin, "Patents and Publicly Funded Research," in *Accessing the Value of Research in the Chemical Sciences*, Chemical Sciences Roundtable, Board on Chemical Sciences and Technology, National Research Council, 1998, accessed at http://www.nap.edu/books/0309061139/html/61.html.

[79] "DOD Technology Advisory Group Says Military Capability Is In Doubt Due to Loss of Electronics Industry," *Manufacturing & Technology News*, May 16, 2003

[80] Pete Engardio, Bruce Einhorn, Manjeet Kripalani, Andy Reinhardt, Bruce Nussbaum, and Peter Burrows, "Outsourcing Innovation," *Business Week Online*, March 21, 2005, accessed at http://www.businessweek.com/magazine/content/05_12/b3925601.htm.

[81] Battelle, "2005 R&D Funding Forecast," *R&D Magazine*, January 2005, accessed at http://www.battelle.org/news/2005rdforecast.pdf.

[82] Ibid.

[83] President's Council of Advisors on Science and Technology, "Sustaining the Nation's Innovation Ecosystems: Report on Information Technology, Manufacturing and Competitiveness," January 2004, accessed at https://www.dodmantech.com/pubs/FINAL_PCAST_IT_Manuf_Report.pdf.

[84] John T. Bennett, "Sega: Pentagon Plan Will Bolster Science and Technology Workforce," *Inside the Pentagon*, April 21, 2005.

Chapter 3

[1] Dana Hicks, "Asian Countries Strengthen Their Research," *Issues in Science and Technology*, Summer 2004, accessed at http://www.issues.org/issues/20.4/realnumbers.html.

[2] National Innovation Initiative, "Innovate America," Council on Competitiveness, December 2004, accessed at http://www.compete.org/pdf/NII_Final_Report.pdf.

[3] William J. Broad, "U.S. Is Losing Its Dominance in the Sciences," *New York Times*, May 3, 2004.

[4] Richard B. Freeman, "Does Globalization of the Scientific/Engineering Workforce Threaten U.S. Economic Leadership?", draft paper prepared for Innovation Policy and the Economy Conference, April 19, 2005, Washington D.C., accessed at http://www.nber.org/~confer/2005/IPEs05/freeman.pdf.

[5] Task Force on Future of Innovation, "The Knowledge Economy: Is the United States Losing Its Competitive Edge?", February 16, 2005, accessed at http://www.futureofinnovation.org/PDF/Benchmarks.pdf.

[6] Freeman, *op. cit.*

[7] National Science Board, Science and Engineering Indicators 2004, Volume 1 (NSB 04-1) and Volume 2 (NSB 04-1A), Arlington, VA: National Science Foundation, accessed at http://www.nsf.gov/sbe/srs/seind04.

[8] NSF Indicators 2004, appendix table 2-33.

[9] "U.S. Slide in World Share Continues as European Union, Asia Pacific Advance," *Science Watch*, July/August 2005, Vol. 16, No. 4, accessed online at http://sciencewatch.com/july-aug2005/sw_july-aug2005_page1.htm.

[10] These data are based on figures from Thompson Scientific National Science Indicators, a database containing publication and citation statistics for more than 100 subfields representing all areas of science.

[11] SCI® is a product of Thompson Scientific. Details of its scope and content can be accessed at http://www.isinet.com.

[12] Robert J. Samuelson, "It's Not a Science Gap (Yet)," *The Washington Post*, August 10, 2005.

[13] Bruce Stokes, "China's High-Tech Challenge," *National Journal*, Vol. 37, No. 31, July 30, 2005, accessed at http://www.aeanet.org/AeACouncils/HUsiqzFY gy.pdf.

[14] Michael L. Marshall, Timothy Coffey, Fred E. Saalfeld, and Rita R. Colwell, "The Science and Engineering Workforce and National Security," *Defense Horizons* 39, Washington, D.C., National Defense University Press, April 2004, accessed at http://www.ndu.edu/ctnsp/defense_horizons/DH39.pdf.

[15] Michael McGeary and Stephen A. Merrill, "Recent Trends in Federal Spending on Scientific and Engineering Research: Impacts on Research Fields and Graduate Training," Appendix A in *Securing America's Industrial Strength*, National Research Council, Washington, D.C.: The National Academies Press, 1999, accessed at http://books.nap.edu/books/0309064481/html/53.html#pagetop.

[16] Stephen A. Merrill, ed., "Trends in Federal Support of Research and Graduate Education," Washington, D.C.: The National Academies Press, 2001.

[17] Elisa Eiseman, Kei Koizumi, and Donna Fossum, "Federal Investment in R&D," RAND Science and Technology Policy Institute, MR-1639.0-OSTP, September 2002, accessed at http://www.rand.org/publications/MR/MR1639.0/MR1639.0.pdf.

[18] President's Council of Advisors on Science and Technology, "Assessing the U.S. R&D Investment: Findings and Proposed Actions," October 16, 2002, accessed at http://www.ostp.gov/PCAST/FINAL%20R&D%20REPORT%20WITH%20LETTERS.pdf.

[19] G. Wayne Clough, Washington, D.C., to John H. Marburger III and E. Floyd Kvamme, Washington, D.C., October 16, 2002, transmitting the panel's final report, "Assessing the U.S. R&D Investment: Findings and Proposed Actions," accessed at http://www.ostp.gov/PCAST/FINAL%20r&d%20REPORT%20WITH%20LETTERS.pdf.

[20] Francis Narin, "Patents and Publicly Funded Research," in Assessing the Value of Research in the Chemical Sciences, Chemical Sciences Roundtable, Board on Chemical Sciences and Technology, National Research Council, 1998, accessed at http://www.nap.edu/books/0309061393/html/61.html.

[21] President's Council of Advisors on Science and Technology, "Assessing the U.S. R&D Investment: Findings and Proposed Actions," October 16, 2002, accessed at http://www.ostp.gov/PCAST/FINAL%20r&d%20REPORT%20WITH%20LETTERS.pdf.

[22] Ibid.

[23] Ibid.

[24] Michael E. Porter and Debra van Opstal, "U.S. Competitiveness 2001: Strengths, Vulnerabilities and Long-Term Priorities," Council on Competitiveness, January 2001, accessed at http://www.compete.org/pdf/competitiveness2001.pdf.

[25] Ibid.

[26] Ibid.

[27] William A. Wulf, "A Disturbing Mosaic," The Bridge, Vol. 35, No. 3, Fall 2005, a publication of the National Academy of Engineering, accessed at http://www.nae.edu/NAE/bridgecom.nsf/weblinks/MKEZ-6GDK3W?OpenDocument.

[28] Ernest H. Preeg, "China's Advanced Technology State," The Washington Times, September 9, 2005, accessed on line at http://www.washingtontimes.com/commentary/20050908-090641-1595r.htm.

[29] Ernest H. Preeg, "The Emerging Chinese Advanced Technology Superstate," MAPI and the Hudson Institute, July 2005.

30 Ronald Kostoff *et al.,* "The Structure and Infrastructure of the Global Nanotechnology Literature," DTIC Technical Report ADA435984, Defense Technical Information Center, 2005, accessed at http://www.onr.navy.mil/sci_tech/special/354/technowatch/docs/nano_dt ic_report_textmin.doc.

31 The Compendix® is a product sold by Engineering Village 2, a web-based information service offering resources in the applied science, technical and engineering fields. Details of its scope and content can be accessed at http://www.engineeringvillage2.org/controller/servlet/Controller

32 Roger Bacon, "*De mirabili potestate Artis et Naturae,*" referenced in the *New Advent,* accessed at http://www.newadvent.org/cathen/13111b.htm.

33 Committee on Advanced Energetic Materials and Manufacturing Technologies, Board on Manufacturing and Engineering Design, Division on Engineering and Physical Sciences, National Research Council of the National Academies, "Advanced Energetic Materials," Washington, D.C.: The National Academies Press, 2004, accessed at http://books.nap.edu/catalog/10918.html.

34 U.S. Department of Commerce, Bureau of Export Administration, Office of Strategic Industries and Economic Security, Strategic Analysis Division, "National Security Assessment of the High Performance Explosives and Explosive Components Industries: A Report for the U.S. Department of the Navy," June 2001.

35 Personal communication cited in the report *Advanced Energetic Materials.*

36 This number is cited in a briefing provided by Robert Kavetsky of the Office of Naval Research.

37 See for example the report by Knezo and references cited therein: Genevieve J. Knezo, "Possible Impacts of Major Counter Terrorism Security Actions on Research, Development, and Higher Education," Congressional Research Service Report for Congress, April 8, 2002, accessed at http://www.aau.edu/research/crsterror.pdf.

38 William A. Wulf, "A Disturbing Mosaic," *op. cit.*

39 Testimony of Norman R. Augustine before the Committee on Education and The Workforce, Subcommittee on 21ˢᵗ Century Competitiveness, U.S. House of Representatives, Hearing on "Challenges to American Competitiveness in Math and Science," May 19, 2005, accessed at http://edworkforce.house.gov/hearings/109th/21stmathscience051905/au gustine.htm.

Chapter 4

[1] Joint Defense Capabilities Study Team, "Joint Defense Capabilities Study: Improving DOD Strategic Planning, Resourcing and Execution to Satisfy Joint Capabilities," Final Report, January 2004, accessed at http://www.acq.osd.mil/actd/articles/JointDefenseCapabilitiesStudyFinal Report_January2004.pdf.

[2] Richard H. Van Atta, Michael J. Lippitz, and Robert L. Bovey, "DOD Technology Management in a Global Technology Environment," IDA Paper P-4017, Alexandria, VA: Institute for Defense Analyses, May 2005.

[3] Darren D. Heusel, *Tinker Take Off*, the official newspaper of Oklahoma City Air Logistics Center, July 12, 2002, accessed at http://www.tinkertakeoff.com/cgi-bin/NewsList.cgi?cat=News&rec=1637&cat=Tinker%2BTake%2BOff.

[4] A good discussion of interdisciplinary research can be found in "Facilitating Interdisciplinary Research," Washington, D.C.: The National Academies Press, 2004, accessed at http://www.nap.edu/openbook/0309094356/html.

[5] Robert Rycroft, "Technology-Based Globalization Indicators: The Centrality of Innovation Network Data," George Washington University, Occasional Paper Series, October 7, 2002, accessed at http://www.gwu.edu/~cistp/PAGES/Tech-BasedGlobIndic_RWR_10.7.02.pdf.

[6] *Ibid.*

[7] Defense Science Board Task Force, "Technology Capabilities of Non-DOD Providers," Office of the Under Secretary of Defense for Acquisition, Technology & Logistics, Washington, D.C., June 2000, accessed at http://www.acq.osd.mil/dsb/reports/technondod.pdf.

[8] National Innovation Initiative, "Innovate America," Council on Competitiveness, December 2004, accessed at http://www.compete.org/pdf/NII_Final_Report.pdf.

[9] John F. Holzrichter, "Attracting and Retaining R&D Talent for Defense," *Physics Today*, April 2001, accessed at http://www.physicstoday.org/pt/vol54/iss-4/p56.html.

[10] Committee on Science, Engineering, and Public Policy, "Facilitating Interdisciplinary Research," Washington, D.C.: The National Academies Press, 2004, accessed at http://www.nap.edu/openbook/0309094356/html.

[11] *Ibid.*

[12] "The Engineer of 2020: Visions of Engineering in the New Century," Washington, D.C.: The National Academies Press, 2004, accessed at http://print.nap.edu/pdf/0309091624/pdf_image/1.pdf.

[13] National Academy of Engineering of the National Academies, "Educating the Engineer of 2020: Adapting Engineering Education to the New Century," Washington, D.C.: The National Academies Press, 2005, accessed at http://print.nap.edu/pdf/030909649/pdf_image/R1.pdf.

[14] American Society of Civil Engineers "Civil Engineering Body of Knowledge for the 21st Century: Preparing the Civil Engineer for the Future," January 12, 2004, accessed at http://www.asce.org/files/pdf/bok/bok_complete.pdf.

[15] "Educating the Engineer of 2020: Adapting Engineering Education to the New Century," op. cit.

[16] "Certificates: A Survey of Our Status and Review of Successful Programs in the U.S. and Canada," a collection of papers presented at the 37th annual meeting of the Council of Graduate Schools in Washington, D.C., December 1997.

[17] Information on the CECD can be found on its website accessed at http://www.cecd.umd.edu.

[18] Data taken from National Academy of Engineering website accessed at http://www.nae.edu.

[19] Office of the Under Secretary of Defense for Acquisition, Technology, and Logistics, "Defense Science Board Task Force on Human Resources Strategy," February 2000.

[20] United States Senate letter of January 6, 2004 to the Honorable Donald Rumsfeld, Secretary of Defense, signed by George V. Voinovich, Susan Collins, John Sessions, Joseph I. Lieberman, and Mike DeWine.

[21] Defense Science Board, "1999 Summer Study Task Force on 21st Century Defense Technology Strategies," Vol. II, March 2000.

[22] Defense Science Board, "Summary of the Defense Science Board Recommendations on DOD Science and Technology Funding," June 1, 2000.

[23] Information on the BENS organization can be accessed at http://www.bens.org.

[24] Naval Research Advisory Committee, "Science and Technology Community in Crisis," May 2002, accessed at http://nrac.onr.navy.mil/webspace/reports/S-n-T_crisis.pdf.

[25] Holzrichter, "Attracting and Retaining R&D Talent," op. cit.

[26] National Academy of Public Administration, "Civilian Workforce 2020: Strategies for Modernizing Human Resources Management in the Department of the Navy," (NAPA 2020 Study), August 18, 2000.

27 Federal Advisory Commission on Consolidation and Conversion of Defense Research and Development Laboratories, *op. cit.*

28 Kenneth Lackie, Jill Dahlberg, Don DeYoung, and Maribel Soto, "Naval Science & Technology and the 'Quiet Crisis'," a Naval Research Laboratory report for the Chief of Naval Research, September 2000.

29 "Civilian Workforce 2020: Strategies for Modernizing Human Resources Management in the Department of the Navy," *op. cit.*

30 Naval Research Advisory Committee, "Science and Technology Community in Crisis," May 2002, accessed at http://nrac.onr.navy.mil/webspace/reports/S-n-T_crisis.pdf.

31 Compensation includes pay, retirement, and benefits such as vacation, holiday, and sick leave.

32 Howard Risher, "Compensating Today's Technical Professional," *Research Technology Management*, January/February 2000.

33 Peter F. Drucker, The Effective Executive, New York, NY, Harper & Row, 1966).

34 Hans Mark, "The Management of Research Institutions," NASA SP 481, 1984.

35 Cited in a speech by the president of the University of Pennsylvania, quoted in "IAST and the Vagelos Laboratories: The Sign and Symbol of a New Research Agenda," accessed at http://www.upenn.edu/almanac/v42/n9/iast.html.

36 Health and Medical Research Strategic Review, "The Virtuous Cycle, Working Together for Health and Medical Research," Final Report, Commonwealth of Australia, May 12, 1999.

37 M. F. Fox, "Publication Productivity Amongst Scientists: A Critical Review," *Social Studies of Science*, Vol. 13, No. 2, May, 1983.

38 P.D. Allison, J.S. Long and T.K. Krauze, "Cumulative Advantage and Inequality in Science," *American Sociological Review*, Vol. 47, No. 5, October 1982, pp. 615-625.

39 Vannevar Bush, Science: The Endless Frontier, Washington, D.C., July 1945.

40 White House Science Council, "Report of the White House Science Council Federal Laboratory Review Panel," May 1983.

41 White House Science Council, Federal Coordinating Committee on Science and Technology Funding Working Group, "Report on Funding Recommendations," May 1984.

42 Kenneth Lackie, Jill Dahlberg, Don DeYoung, and Maribel Soto, *op. cit.*

43 Kenneth Lackie, Jill Dahlberg, Don DeYoung, and Maribel Soto, *op. cit.*

44 Peter F. Drucker, Management: Tasks, Responsibilities, Practices, New York, NY, 1973/74.

[45] Office of the Director of Navy Laboratories, "A Survey of Reports on the Management of Research and Development in the Postwar Navy," December 16, 1987.

[46] Department of Defense, Task Force 97 Action Group, "Review of Defense Laboratories: Progress Report and Preliminary Recommendations," September 1961.

[47] Bureau of the Budget, "Report to the President on Government Contracting for Research and Development," (Bell Report), April 30, 1962.

[48] Office of the Director of Defense Research and Engineering, "Report of the Defense Science Board on Government In-House Laboratories," September 6, 1962.

[49] "Report to the President and the Secretary of Defense on the Department of Defense," July 1, 1970.

[50] Booz, Allen & Hamilton, "Review of Navy R&D Management 1946-1973," June 1, 1976.

[51] Report of the Defense Science Board 1987 Summer Study on Technology Base Management, December 1987.

[52] ADM Vernon Clark, Naval message no. 271955Z of March 2003.

[53] A. van de Vliat, "Lest We Forget," *Management Today*, January 1995; *HR Magazine*, "Losing Corporate Memory, Bit by Bit," May 1996; Annie Brooking, *Intellectual Capital*, International Thomson Business Press, Andover, England, 1999; K. Kreiner and M. Schultz, "Informal Collaboration in R&D: The Formation of Networks," *Organizational Studies*, 1993; W.E. Baker, "Bloodletting and Downsizing," *Executive Excellence*, May 1996.

[54] L. Ryne and M. Teagarden, "Technology-based competitive strategy: An empirical test of an integrative model," *Journal of High Technology Management Research*, Fall 1997, Vol. 8, No. 2, p. 187.

[55] Center for Innovation Management Studies, "Now Comes the Capability Crisis—Is Your Company Ready?", Technology Management Report, Spring 2005, NC State University, Raleigh, NC, accessed at http://cims.ncsu.edu/documents/spring05.pdf.

[56] David W. DeLong, Lost Knowledge: Confronting the Threat of an Aging Workforce, New York, NY, 2004.

[57] Dorothy Leonard and Walter Swap, "Deep Smarts," *Harvard Business Review*, September 2004, Vol. 82, Issue 9, pp. 88-97.

[58] Committee on the Future of the U.S. Aerospace Infrastructure and Aerospace Engineering Disciplines to Meet the Needs of the Air Force and the Department of Defense, Air Force Science and Technology Board, Division on Engineering and Physical Sciences, National Research Council, "Review of the Future of the U.S. Aerospace Infrastructure and Aerospace Engineering Disciplines to Meet the Needs

of the Air Force and the Department of Defense," Washington D.C.: The National Academies Press, 2001, accessed at http://www.nap.edu/books/0309076064/html.

[59] Naval Research Advisory Committee, "Science and Technology Community in Crisis," May 2002, accessed at http://nrac.onr.navy.mil/webspace/reports/S-n-T_crisis.pdf.

[60] Panel on Presidentially Appointed Scientists and Engineers, Committee on Science, Engineering, and Public Policy, National Academy of Sciences, National Academy of Engineering, Institute of Medicine, "Science and Technology Leadership in American Government: Ensuring the Best Presidential Appointments," Washington, D.C.: The National Academies Press, 1992, accessed at http://books.nap.edu/books/0309047277/html/R1.html .

[61] Megan Scully, "Senator Seeks Route for England to Take Pentagon Post," *GOVEXEC.COM*, June 23, 2005, accessed at http://www.governmentexecutive.com/story_page.cfm?articleid=31583&sid=21.

[62] Cheryl Y. Marcum, Lauren R. Sager Weinstein, Susan D. Hosek, and Harry J. Thie, "Department of Defense Political Appointments: Positions and Process," RAND National Defense Research Institute, 2001.

[63] Defense Science Board Task Force, "Human Resources Strategy," Office of the Under Secretary of Defense for Acquisition, Technology, and Logistics, February 2000.

[64] Robert Klitdaard and Paul C. Light, editors, "High-Performance Government: Structure, Leadership, Incentives," Pardee Grand Graduate School, 2005, accessed at http://www.rand.org/publications/MG/MG256.

[65] Senate Report 108-260 to accompany S. 2400, May 11, 2004.

[66] Federal Laboratory Review Panel, "Report of the White House Science Council," May 1983.

[67] Department of Defense, Task Force 97 Action Group, "Review of Defense Laboratories: Progress Report and Preliminary Recommendations," September 1961.

[68] Booz, Allen & Hamilton, "Review of Navy R&D Management 1946-1973," June 1, 1976.

[69] Defense Science Board Task Force, "In-House Laboratories," October 31, 1966.

[70] The dual executive instruction was cancelled by the Principal Deputy ASN (RDA) in a memorandum of June 15, 1995, Subject: Cancellation of Secretary of the Navy (SECNAV) Instructions.

[71] Quoted in a briefing by LtGen Stephen Plummer entitled "AF Scientists and Functional Manager Update," given at an Air Force summit on S&E workforce issues held November 5-7, 2002.

[72] NRAC, "Science and Technology Community in Crisis."

[73] Office of the Chief Scientist of the Air Force, "Science and Technology Workforce For the 21st Century: A Report prepared for the Acting Secretary and Chief of Staff of the Air Force," Washington, D.C., July 1999.

[74] Air Force Scientific Advisory Board, "Science and Technology and the Air Force Vision: Achieving a More Effective Science and Technology Program," SAB-TR-00-02, May 2001.

[75] Air Force Science and Technology Board, "Effectiveness of Air Force Science and Technology Program Changes," National Research Council of the National Academies, 2003, accessed at http://www.nap.edu/books/030908895X/html.

Chapter 5

[1] Commentary, *Defense News*, May 9, 2005.

[2] "US Used New Missile in Iraq: Rumsfeld," *Sydney Morning Herald*, May 15, 2003, accessed at http://www.smh.com.au/articles/2003/05/15/1052885324203.html.

[3] For an early example of an effort to quantify the return from the DON's investment in basic research, see Bruce S. Old, "Return on Investment in Basic Research—Exploring a Methodology," *The Bridge*, Vol. 12, No. 1, National Academy of Engineering, Spring 1982.

[4] Tom Pelsoci, "Benefit-to-Cost Analysis of NUWC ILIR Portfolio," Delta Research Corporation, December, 2005.

[5] William Nordhaus, "The Economic Consequences of a War with Iraq (Revised)," Yale University, October 29, 2002, accessed at http://www.econ.yale.edu/~nordhaus/iraq.pdf.

[6] John F. Holzrichter, "Attracting and Retaining R&D Talent for Defense," *Physics Today*, April 2001, accessed at http://www.physicstoday.org/pt/vol54/iss-4/p56.html.

[7] *Ibid.*

[8] Peter Navarro and Aron Spencer, "September 11, 2001: Assessing the Costs of Terrorism," *The Milken Institute Review*, Fourth Quarter 2001.

[9] Gary Becker and Kevin Murphy, "Prosperity Will Rise Out of the Ashes," *Wall Street Journal*, October 29, 2001, cited in "The Economic Costs of Terrorism," a study for the Joint Economic Committee of the United States Congress, May 2002.

[10] Navarro and Spencer 2001, *op. cit.*

[11] International Monetary Fund, Chapter 11, "How Has September 11 Influenced the Global Economy," World Economic Outlook, December 2001, cited in "The Economic Costs of Terrorism," a study for the Joint Economic Committee of the United States Congress, May 2002.

[12] Atul Gawande, "Casualties of War—Military Care for the Wounded from Iraq and Afghanistan," *New England Journal of Medicine*, Vol. 351, No. 24, December 9, 2004, accessed at http://content.nejm.org/cgi/content/full/351/24/2471#R1.

[13] A running total of casualties for both the Iraq and Afghanistan conflicts can be accessed at http://www.defenselink.mil/news/casualty.pdf.

[14] "US Senate Approves Emergency Vets Health Care Funds," Reuters, June 29, 2005.

[15] Statement of the Honorable R. James Nicholson before the House Veterans Affairs Committee, June 30, 2005, accessed at http://veterans.house.gov/hearings/schedule109/jun05/6-30-05/nicholson.pdf.

Chapter 6

[1] The Wexford Group International, "Workforce Gap Analysis for the Office of Naval Research (ONR) Science and Technology (S&T) Revitalization Initiative," March 9, 2004.

[2] An S&T worker is defined as an individual scientists or engineer who charges at least 50 percent of his time to an S&T project.

[3] The concept of an "Esteemed Fellow" is an honorary designation bestowed upon a small number of individuals by NRE scientific colleagues. These individuals are recognized experts in their technical field and are at the ST/SL or equivalent in civilian grade. A very small percentage of S&Es attain the designation of "Esteemed Fellows."

[4] American Society of Civil Engineers, "Civil Engineering Body of Knowledge for the 21st Century: Preparing the Civil Engineer for the Future," January 12, 2004, accessed at http://www.asce.org/files/pdf/bok/bok_complete.pdf.

[5] "Educating the Engineer of 2020: Adapting Engineering Education to the New Century," *op. cit.*

[6] BEST reports can be accessed at http://www.bestworkforce.org/publications.htm.

[7] In order to be counted as a member of the S&T workforce, an individual S&E must charge at least half of his total time, measured on an annual basis, to projects funded by either BA 1, BA 2 or BA 3.

[8] Timothy Coffey, Jill Dahlburg and Eli Zimet, "The S&T Innovation Conundrum," Center for Technology and National Security Policy, National Defense University, accessed at http://www.ndu.edu/ctnsp/Def_Tech/DTP%2017%20S&T%20Innovation%20Conundrum.pdf.

[9] Quadrennial Defense Review Report, Department of Defense, September 30, 2001, accessed at http://www.defenselink.mil/pubs/qdr2001.pdf.

[10] Office of the Under Secretary of Defense for Acquisition and Technology, "Report of the Defense Science Board Task Force on Defense Science and Technology Base for the 21st Century," Washington, D.C., June 30, 1998, accessed at http://www.acq.osd.mil/dsb/reports/sandt21.pdf; Office of the Under Secretary of Defense for Acquisition, Technology and Logistics, "Defense Science Board 2001 Summer Study on Defense Science and Technology," Washington D.C., May 2002, accessed at http://www.acq.osd.mil/dsb/reports/sandt.pdf.

[11] April 2002 letter from several Senate Armed Services Committee members to Committee Chairman Carl Levin and Ranking Member John Warner.

[12] House Armed Services Committee report language in the fiscal year 2003 Defense Authorization Committee report, May 3, 2002.

[13] Senate Armed Services Committee report language in the fiscal year 2003 Defense Authorization Committee report, May 15, 2002.

[14] Rear Admiral Jay M. Cohen, USN, Chief of Naval Research, Testimony before the Senate Armed Services Committee, Subcommittee on Emerging Threats and Capabilities, Defense Authorization Hearing, April 10, 2002, accessed at http://armed-services.senate.gov/statemnt/2002/April/Cohen.pdf.

[15] Naval Research Advisory Committee, "Science and Technology Community in Crisis," May 2002, accessed at http://nrac.onr.navy.mil/webspace/reports/S-n-T_crisis.pdf.

[16] National Science Foundation Facility Plan, September 2005, accessed at http://www.nsf.gov/pubs/2005/nsf05058/nsf05058.pdf.

[17] Details on the DURIP can be found in the FY 2006 program announcement, accessed at http://www.onr.navy.mil/sci_tech/industrial/363/docs/050519_durip_2006.doc

[18] "U.S. Companies Failing to Transfer Critical Knowledge," Management Issues, May 10, 2005, accessed at http://www.management-issues.com/display_page.asp?section=research&id=2118.

[19] Canada School of Public Service, "A Primer on Knowledge Management in Public Service: A Working Paper Prepared for the Canadian Centre for Management Development's Action-Research Roundtable on the Learning Organization," April 4, 1999, accessed at http://www.myschool-monecole.gc.ca/research/publications/pdfs/LO-Primer-REV.PDF.

[20] Senate Report 108-260, National Defense Authorization Act for Fiscal Year 2005 Report [to accompany S. 2400] on Authorizing Appropriations For Fiscal Year 2005 for Military Activities of the Department of Defense, for Military Construction, and for Defense Activities of the Department of Energy, to Prescribe Personnel Strengths For Such Fiscal Year for the Armed Forces, and for Other Purposes Together With Additional Views, accessed at http://thomas.loc.gov/cgi-bin/cpquery.

[21] Federal Advisory Commission on Consolidation and Conversion of Defense Research and Development Laboratories, *op. cit.*

[22] William J. Moran, VADM USN (Ret), interview by R.L. Hansen for Booz, Allen, Hamilton R&D management study, BA-5, 1975. Cited in "You Can't Run a Laboratory Like a Ship," a white paper prepared for the Office of the Director of Navy Laboratories by A. B. Christman, Historian of Navy Laboratories.

[23] Discussion of the Air Force situation is based on personal communication of June 17, 2005 from Donald Lamberson (Major General U.S. Air Force Retired) to Michael Marshall; personal communication of June 21, 2005 from Robert Rankine (Major General U.S. Air Force Retired) to Michael Marshall.

[24] Committee on Prospering in the Global Economy of the 21st Century: An Agenda for American Science and Technology, Committee on Science, Engineering and Public Policy, "Rising Above the Gathering Storm: Energizing and Employing America for a Brighter Economic Future," Washington, D.C.: The National Academies Press, 2005, accessed at http://www.nap.edu/catalog/11463.html.

[25] *Saturday Review*, April 15, 1978.